# SpringerBriefs in Computer Science

SpringerBriefs present concise summaries of cutting-edge research and practical applications across a wide spectrum of fields. Featuring compact volumes of 50 to 125 pages, the series covers a range of content from professional to academic.

Typical topics might include:

- A timely report of state-of-the art analytical techniques
- A bridge between new research results, as published in journal articles, and a contextual literature review
- A snapshot of a hot or emerging topic
- An in-depth case study or clinical example
- A presentation of core concepts that students must understand in order to make independent contributions

Briefs allow authors to present their ideas and readers to absorb them with minimal time investment. Briefs will be published as part of Springer's eBook collection, with millions of users worldwide. In addition, Briefs will be available for individual print and electronic purchase. Briefs are characterized by fast, global electronic dissemination, standard publishing contracts, easy-to-use manuscript preparation and formatting guidelines, and expedited production schedules. We aim for publication 8–12 weeks after acceptance. Both solicited and unsolicited manuscripts are considered for publication in this series.

\*\*Indexing: This series is indexed in Scopus, Ei-Compendex, and zbMATH \*\*

Hao Kong • Jiadi Yu • Xuemin (Sherman) Shen

# Human Reconstruction Using mmWave Technology

 Springer

Hao Kong ⓘ
Shanghai, China

Jiadi Yu
Shanghai, China

Xuemin (Sherman) Shen
Waterloo, ON, Canada

ISSN 2191-5768                ISSN 2191-5776    (electronic)
SpringerBriefs in Computer Science
ISBN 978-3-032-01148-0       ISBN 978-3-032-01149-7    (eBook)
https://doi.org/10.1007/978-3-032-01149-7

© The Editor(s) (if applicable) and The Author(s), under exclusive license to Springer Nature Switzerland AG 2026

This work is subject to copyright. All rights are solely and exclusively licensed by the Publisher, whether the whole or part of the material is concerned, specifically the rights of translation, reprinting, reuse of illustrations, recitation, broadcasting, reproduction on microfilms or in any other physical way, and transmission or information storage and retrieval, electronic adaptation, computer software, or by similar or dissimilar methodology now known or hereafter developed.
The use of general descriptive names, registered names, trademarks, service marks, etc. in this publication does not imply, even in the absence of a specific statement, that such names are exempt from the relevant protective laws and regulations and therefore free for general use.
The publisher, the authors and the editors are safe to assume that the advice and information in this book are believed to be true and accurate at the date of publication. Neither the publisher nor the authors or the editors give a warranty, expressed or implied, with respect to the material contained herein or for any errors or omissions that may have been made. The publisher remains neutral with regard to jurisdictional claims in published maps and institutional affiliations.

This Springer imprint is published by the registered company Springer Nature Switzerland AG
The registered company address is: Gewerbestrasse 11, 6330 Cham, Switzerland

If disposing of this product, please recycle the paper.

# Preface

Human and the cyberworld are moving towards high synchronization as more Internet of Things (IoT) devices are integrated in smart homes. By reconstructing human postures, facial expressions, and hand gestures in the cyberworld, a virtual human avatar can map the state of human into the physical world and deeply integrate reality and virtuality. The process strongly energizes emerging applications such as virtual reality (VR), augmented reality (AR), meta-universe, immersive games, etc. The methods of human reconstruction usually rely on wearables and cameras. However, wearables require users to actively wear specialized equipment, introducing intrusive user experiences and resulting in high costs. Vision approaches depend on the lighting conditions of environments and suffer from privacy leakage concerns, which are increasingly getting attention nowadays.

Toward a more robust and privacy-preserving sensing manner, radio frequency (RF) signals have been exploited for ubiquitous sensing beyond only communications. Among various RF sensing methods, millimeter wave (mmWave) is considered one of the promising solutions. Sensing with mmWave is robust to different lighting conditions and works in a privacy-preserving manner. mmWave can penetrate some obstacles and expand the use cases of vision-based approaches. Besides, as a contactless sensing manner, it provides nonintrusive user experiences compared with wearable devices. Along with its properties, the high applicability motivates us to leverage mmWave to build nonintrusive human reconstruction, ranging from coarse-grained body postures to fine-grained facial expressions and hand gestures.

In this monograph, we introduce human reconstruction technologies in smart homes leveraging mmWave signals. It begins by introducing the overview of human reconstruction and the development of mmWave sensing technology in Chap. 1. Chapter 2 proposes a mmWave sensing-based human posture reconstruction approach, $m^3Track$, which exploits mmWave signals to sense and track multiple users' postures as they move, walk, or sit. Chapter 3 presents *mm3DFace*, a facial expression reconstruction system that reconstructs 3D human faces and continuously exhibits facial expressions using mmWave. Chapter 4 describes a hand gesture reconstruction system, *mmHand*, which utilizes mmWave signals to generate 3D hand skeletons and reconstruct hand meshes. A thorough investigation of state-of-the-art research work is presented in Chap. 5 covering human reconstruction, mmWave sensing, and mmWave-based human reconstruction. Finally, conclusions of the monograph and the direction of future research are given in Chap. 6.

We would like to thank the members of the research group at Shanghai Jiao Tong University, and BBCR group at the University of Waterloo, for their valuable suggestions and support. Special thanks to the staff at Springer Nature, particularly, for their invaluable help throughout the publication preparation process.

Shanghai, China  Hao Kong
Shanghai, China  Jiadi Yu
Waterloo, ON, Canada  Xuemin (Sherman) Shen

# Declarations

**Competing Interests** The authors have no competing interests to declare that are relevant to the content of this manuscript.

# Contents

**1 Introduction** .................................................... 1
   1.1 An Overview of Human Reconstruction ........................... 1
   1.2 mmWave Sensing Technology ..................................... 2
      1.2.1 Brief Introduction of mmWave Sensing .................... 2
      1.2.2 Techniques of mmWave FMCW Radar ......................... 3
      1.2.3 mmWave Sensing Applications ............................. 8
   1.3 Organization of Monograph ..................................... 10
   References ......................................................... 11

**2 mmWave-based Human Posture Reconstruction** ....................... 13
   2.1 Introduction .................................................. 13
   2.2 System Overview ............................................... 15
   2.3 Multi-User Detection and Separation ........................... 16
      2.3.1 User Detection on mmWave Signals ........................ 16
      2.3.2 User Separation on mmWave Signals ....................... 17
   2.4 Single-User Posture Reconstruction ............................ 20
      2.4.1 Posture Feature Representation .......................... 20
      2.4.2 3D Posture Reconstruction ............................... 21
   2.5 Multi-User 3D Posture Tracking ................................ 24
      2.5.1 Posture Mapping with Point Cloud ........................ 25
      2.5.2 3D Posture Tracking ..................................... 26
   2.6 Evaluation .................................................... 28
      2.6.1 Evaluation Setup ........................................ 28
      2.6.2 Overall Performance ..................................... 29
      2.6.3 Quantitative Results .................................... 30
      2.6.4 Performance in Different Environments ................... 32
      2.6.5 Performance in Occluded Scenarios ....................... 33
      2.6.6 Performance of User Joining and Leaving ................. 34
      2.6.7 Comparison with SOTA Systems ............................ 35
      2.6.8 Localization Performance ................................ 36
      2.6.9 Impact of Distance between Users ........................ 37

|  |  | 2.6.10 | Impact of Distance to Radar | 39 |
|---|---|---|---|---|
|  | 2.7 | Summary | | 39 |
|  | References | | | 39 |

## 3 mmWave-based Facial Expression Reconstruction — 41
- 3.1 Introduction — 41
- 3.2 System Overview — 44
- 3.3 mmWave Signal Pre-processing — 45
- 3.4 Facial Feature Extraction — 46
  - 3.4.1 Triple-Loss-Embedding-based Facial Geometric Feature Extraction — 48
- 3.5 3D Facial Reconstruction — 50
  - 3.5.1 Facial Shape Reconstruction — 51
  - 3.5.2 Facial Expression Reconstruction — 54
  - 3.5.3 3D Avatar Generation — 56
- 3.6 Evaluation — 58
  - 3.6.1 Evaluation Setup — 58
  - 3.6.2 Overall Performance — 60
  - 3.6.3 Comparison with Existing Method — 63
  - 3.6.4 Facial Expression Recognition — 64
  - 3.6.5 Impact of Mask — 65
  - 3.6.6 Impact of Distance — 66
  - 3.6.7 Impact of Orientation — 68
  - 3.6.8 Impact of Background Environment — 68
  - 3.6.9 Time Consumption — 70
- 3.7 Summary — 71
- References — 72

## 4 mmWave-based Hand Gesture Reconstruction — 75
- 4.1 Introduction — 75
- 4.2 System Overview — 78
- 4.3 Signal Pre-Processing — 79
- 4.4 Hand Joint Regression — 80
  - 4.4.1 Hand Feature Extraction — 82
  - 4.4.2 Regressing 3D Hand Joints Based on a Combined Loss — 84
- 4.5 Mesh Reconstruction — 85
- 4.6 Evaluation — 87
  - 4.6.1 Evaluation Setup — 87
  - 4.6.2 Overall Performance — 89
  - 4.6.3 Comparison with Existing Methods — 92
  - 4.6.4 Impact of Distance — 93
  - 4.6.5 Impact of Angle — 94
  - 4.6.6 Impact of Human Body — 95
  - 4.6.7 Impact of Gloves — 96
  - 4.6.8 Impact of Handheld Object — 97
  - 4.6.9 Impact of Environment — 98

|  |  | 4.6.10 Impact of Obstacle | 98 |
|---|---|---|---|
|  | 4.7 | Summary | 99 |
|  | References | | 100 |
| **5** | **State-of-Art Research** | | **103** |
|  | 5.1 | Human Reconstruction Research | 103 |
|  |  | 5.1.1 Vision-based Approaches | 103 |
|  |  | 5.1.2 Wearable Device-Based Approaches | 104 |
|  | 5.2 | mmWave Sensing Research | 105 |
|  |  | 5.2.1 mmWave Sensing in Autonomous Driving | 106 |
|  |  | 5.2.2 mmWave Sensing in Smart Homes | 106 |
|  |  | 5.2.3 mmWave Sensing in Industrial Manufacture | 107 |
|  | 5.3 | mmWave-based Human Reconstruction | 108 |
|  |  | 5.3.1 mmWave-based Pose Estimation | 108 |
|  |  | 5.3.2 mmWave-based Gesture Recognition | 109 |
|  | 5.4 | Summary of Existing Research | 110 |
|  | References | | 111 |
| **6** | **Conclusions and Future Research Directions** | | **119** |
|  | 6.1 | Conclusions | 119 |
|  | 6.2 | Future Research Directions | 120 |

# Acronyms

| | |
|---|---|
| mmWave | millimeter wave |
| IoT | Internet of Things |
| VR | virtual reality |
| AR | augmented reality |
| FMCW | frequency-modulated continuous-wave |
| TI | Texas Instruments |
| NLOS | non-line-of-sight |
| RFID | radio frequency identification |
| IF | intermediate frequency |
| FFT | fast Fourier transform |
| AoA | angle-of-arrival |
| SLAM | simultaneous localization and mapping |
| LSTM | long short-term memory |
| COTS | commercial off-the-shelf |
| MVDR | minimum variance distortionless response |
| MIMO | multiple-input and multiple-output |
| TDM | time-division multiplexing |
| CNNs | convolutional neural networks |
| CFAR | constant false alarm rate |
| EKF | extended Kalman filter |
| SOTA | state-of-the-art |
| CDF | cumulative distribution function |
| PCA | principal component analysis |
| MAE | mean absolute error |
| NME | normalized mean error |
| MANO | hand model with articulated and non-rigid deformations |
| RC | radar cube |
| TGAP | three-dimensional global average pooling |
| TGMP | three-dimensional global max pooling |
| GMP | global max pooling |
| GAP | global average pooling |

| | |
|---|---|
| SMPL | skinned multi-person linear model |
| MPJPE | mean per joint position error |
| AUC | area under the curve |
| GCNs | graph convolutional networks |
| IMUs | inertial measurement units |
| EMG | electromyography |
| KPF | knitted piezoresistive fabric |
| MSG | multi-scale grouping module |
| CRFNet | CameraRadarFusionNet |
| SAF | spatial attention fusion |
| RCS | radar cross-section |
| ToA | time-of-arrival |
| NLP | natural language processing |
| ISAC | integrated sensing and communication |
| LLMs | large language models |

# Chapter 1
# Introduction

**Abstract** In this chapter, we first give a brief introduction of human reconstruction, covering the core concept, usage scenarios and commonly used approaches. Then, we give an overview of millimeter wave (mmWave) sensing technologies, and also present the emerging applications. Finally, we describe the organization of this monograph.

**Keywords** Human reconstruction · Wireless sensing · mmWave technology · Frequency-modulated continuous wave · Internet of Things · Smart homes

## 1.1 An Overview of Human Reconstruction

Human reconstruction technology refers to the process of digitally reconstructing various aspects of human morphology and movement, such as body poses, hand gestures, and facial expressions. By exploiting the sensor data that captures human body torso or body parts, human reconstruction technology generates accurate, real-time digital representations of daily activities. In today's proliferating Internet of Things (IoT) scenarios, human reconstruction serves as a foundational technology for various applications in smart environments.

Human reconstruction in smart homes enables touchless and intuitive control of IoT appliances. Besides interaction, it can also assist smart home healthcare monitoring through realistically reconstructing human body, which can serve elderly fall detection, rehabilitation monitoring, identifying irregular movements, etc. Moreover, in the emerging market of virtual reality (VR) and augmented reality (AR), human reconstruction also plays an important role. The technology bridges the physical and virtual worlds by accurately replicating user movements in digital spaces, which enhances immersion in VR gaming, virtual meetings, and metaverse environments. As human reconstruction technology evolves, the integration with IoT systems leads to increasingly personalized experiences. Emerging advancements such as AI-driven predictive analytics and physics-based modeling improve reconstruction accuracy and efficiency. Therefore, the convergence of human reconstruction technology of extensive application scenarios will drive innovation in areas like smart cities, healthcare, education, and entertainment, profoundly transforming how humans interact with intelligent systems.

Existing human reconstruction approaches, while effective in many scenarios, are constrained by several limitations that hinder their broader applicability. Vision-based methods have been widely adopted in capturing human activities by various cameras for human reconstruction [1]. However, these methods are highly sensitive to environmental illumination conditions. For instance, low-light environments or overly bright scenes can significantly degrade their accuracy and reliability. Moreover, these approaches struggle in occlusion scenarios where parts of the human body are blocked by objects or other body parts, leading to incomplete reconstruction. A further concern with vision-based methods is the capture of detailed visual information about users and their environments, which can expose sensitive data and lead to privacy concerns, especially in home or workplace settings.

Wearable technologies, such as motion capture suits, sensor-equipped gloves, and smart bands, present another category of human reconstruction approaches [2]. While these methods offer robustness and precision, they introduce several challenges that limit their usability. Users are required to actively wear devices attached to their bodies, which can feel intrusive and burdensome, particularly for long-term or everyday use. This intrusiveness can reduce user acceptance and make the technology less appealing for casual or non-specialist applications. Furthermore, many wearable devices are expensive due to the high cost of advanced sensors and materials, which can create financial barriers to broader adoption in consumer or large-scale industrial contexts.

Therefore, the limitations underscore the need for innovative approaches to human reconstruction. The solutions should operate effectively in diverse environmental conditions, handle occlusion scenarios, and prioritize user privacy without requiring intrusive or costly equipment.

## 1.2 mmWave Sensing Technology

### 1.2.1 Brief Introduction of mmWave Sensing

In today's IoT environments, sensing technology has advanced significantly, enabling the measurement of diverse physical, environmental, and biological phenomena through specialized sensors. By analyzing these measurements, it facilitates understanding and interaction with the real world. A wide range of sensing technologies, especially wireless signals, have been widely utilized in IoT applications. Among them, mmWave radar sensing stands out as a highly promising solution.

mmWave radar sensing technology utilizes modulated signals to extract environmental information. Operating in the frequency range of 30–300 GHz, mmWave was initially developed to support high-speed, ultra-reliable, and low-latency wireless communications [3–5]. Its unique features, including wide bandwidth, short wavelength, and compact antenna arrays, have enabled sensing capabilities that extend beyond communication applications. A standard mmWave radar comprises transmit and receive antenna arrays. The transmit array emits modulated mmWave signals, which propagate through the environment, reflect off objects, and are captured by the receive array. By analyzing the reflected signals, the radar extracts spatial and temporal details about the environment, enabling passive and contactless sensing. This capability is largely powered by frequency-modulated continuous-wave (FMCW) techniques, which underpin several commercial mmWave radars, such as those from Texas Instruments (TI) [6].

mmWave radar sensing offers several advantages. Unlike sensors that require physical contact or integration, its contactless nature ensures nonintrusive user experience while reducing deployment costs. It performs reliably in adverse weather and some non-line-of-sight (NLOS) conditions, overcoming the limitations of cameras and lidars in complex scenarios. Because of the short wavelengths, mmWave radars achieve higher spatial resolution compared to technologies like WiFi, radio frequency identification (RFID), and acoustic sensors. Additionally, as privacy concerns surrounding vision-based systems grow, mmWave radars provide a more privacy-preserving alternative.

These benefits have positioned mmWave radars as pivotal enablers for diverse applications, extending well beyond traditional radar use cases. In autonomous vehicles, mmWave radars play a key role in obstacle detection and motion estimation, contributing to advancements in autonomous driving technologies. In smart homes, they support activity and gesture recognition, enabling "in-air" human-computer interaction for VR applications. In manufacturing, they serve as cost-effective sensors for high-resolution physical measurements. The growing demand for these applications has demonstrated the significance of mmWave sensing in IoT environments, which fuels a surge of system design over the past decade.

### 1.2.2 Techniques of mmWave FMCW Radar

**FMCW Technique** FMCW techniques utilize modulated pulsed waves to sense environments and extract objective information. Their millimeter-level wavelengths enable high-resolution range and velocity estimation, allowing the differentiation of small distances. Additionally, the signal processing after mixing occurs at low frequencies, which significantly simplifies the design of processing circuits. These advantages have led to widespread adoption of FMCW techniques in various applications, making them a dominant choice for mmWave radars.

FMCW techniques enhance the capabilities of traditional continuous wave radars by enabling the detection of close objects with the minimum measurable distance. Using FMCW, radars can estimate both the distance and relative velocity of target objects. The technique involves modulating transmitted mmWave signals into

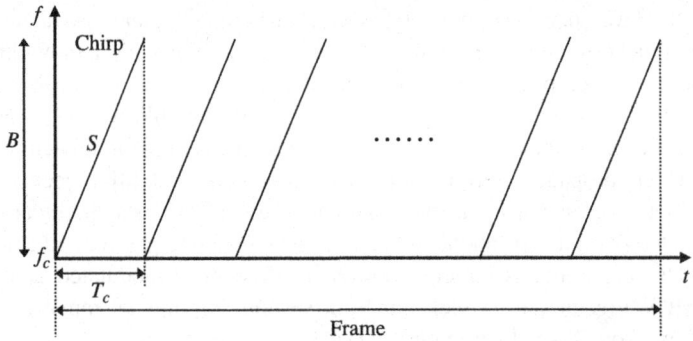

**Fig. 1.1** An example of a chirp frame, where each chirp has a bandwidth $B$, duration $T_c$, slope $S$, and starting start frequency $f_c$

continuous waves with linearly varying frequencies, a process referred to as linear frequency modulation or chirping. The frequency of a chirp is expressed as:

$$f(t) = St + f_c, \tag{1.1}$$

where $f(t)$ represents the frequency, $S$ is the slope of the chirp, $t$ denotes time, and $f_c$ is the starting frequency. The slope $S$ of the chirp is calculated as:

$$S = \frac{B}{T_c}, \tag{1.2}$$

where $B$ is the bandwidth of the chirp, and $T_c$ is its duration. Typically, mmWave radars transmit a sequence of equally spaced chirp signals as a frame for sensing, as illustrated in Fig. 1.1.

A mixer is used to combine the transmitted and received signals, producing a new signal. For two sine signals denoted as $x_1$ and $x_2$, the signals are expressed as:

$$x_1 = \sin(\omega_1 t + \phi_1), \tag{1.3}$$

$$x_2 = \sin(\omega_2 t + \phi_2), \tag{1.4}$$

where $\omega_i$ and $\phi_i$ represent the angular velocity and initial phase of the $i$-th signal, respectively. The output signal from the mixer is given by:

$$x_{\text{out}} = \sin((\omega_1 - \omega_2)t + (\phi_1 - \phi_2)), \tag{1.5}$$

which is referred to as the intermediate frequency (IF) signal [7]. Using the IF signal, mmWave radars perform further data processing to estimate various physical quantities, such as the range and velocity of objects. This capability forms the foundation of basic sensing functions in FMCW radars.

## 1.2 mmWave Sensing Technology

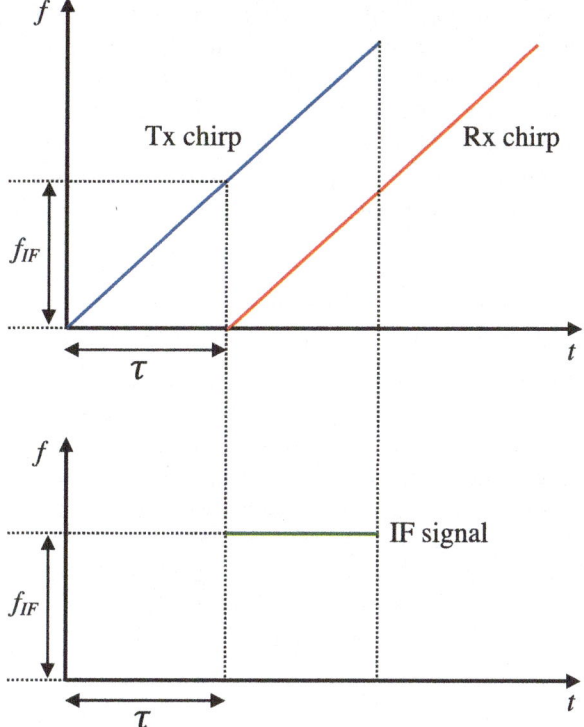

**Fig. 1.2** The Tx signal and Rx signal are mixed to generate the IF signal, which has a delay $\tau$ and constant frequency $f_{IF}$

**Range Estimation** Range estimation is one of the basic functions of mmWave radars. Assume an object is located at a range $d$ from a radar. As shown in Fig. 1.2, a transmitted chirp signal is reflected by the object and arrives at the receive antennas with a time delay. The frequency of the IF signal is the difference between the transmitted signal and the received signal. Hence, we have:

$$\tau = \frac{2d}{c}, \tag{1.6}$$

$$S = \frac{f_{IF}}{\tau}, \tag{1.7}$$

where $\tau$ is the delay between the transmitted and received signals, $d$ is the object's range, $c$ is the speed of light, $S$ is the slope of the chirp, and $f_{IF}$ is the frequency of the IF signal. Based on these formulas, the range of the object can be estimated as:

$$d = \frac{cT_c f_{IF}}{2B}, \tag{1.8}$$

where $T_c$ is the chirp's duration and $B$ is the chirp's bandwidth.

When multiple objects are at different ranges, each object reflects the transmitted chirp signal and produces a reflected chirp signal with a delay proportional to its range. Consequently, the IF signal consists of multiple tones, each with a constant frequency corresponding to a specific target. For such IF signals, a fast Fourier transform (FFT) [8] is employed to separate the tones in the frequency spectrum, a process known as Range-FFT. Each peak in the frequency spectrum represents an object at a specific range, enabling the estimation of the ranges of multiple objects using FMCW techniques.

As a fundamental function of mmWave radars, range estimation describes the relative distances between radar and objects. This capability is essential for distance-based applications and supports the further processing of mmWave signals.

**Velocity Estimation** Velocity is critical information that describes the instantaneous motion state of sensed objects. To estimate the velocity of a moving object, a radar transmits two chirp signals separated by $T_c$ and receives the signals reflected by the object. Since $T_c$ is typically measured in milliseconds, the object's movement is smaller than the range resolution, resulting in a single peak in Range-FFT. However, the object's motion introduces a phase difference between the reflected chirp signals. The phase of a chirp signal can be expressed as $\phi = (4\pi d)/\lambda$, where $\lambda$ is the wavelength. The object's displacement can be denoted as $\Delta d = vT_c$, leading to the velocity estimation formula:

$$v = \frac{\lambda \Delta \phi}{4\pi T_c}, \tag{1.9}$$

where $v$ is the velocity, and $\Delta \phi$ is the phase difference between the two reflected chirp signals.

If multiple objects are moving simultaneously within the same range, straightforward phase difference estimation fails because the phase differences overlap among multiple objects. To address this, a radar transmits a chirp frame with $N_c$ equally spaced chirps and performs Range-FFT on the reflected chirp frame. Multiple objects generate $N_c$ peaks with the same range in the frequency spectrum, each having a distinct phase due to their movement, as illustrated in Fig. 1.3. A second FFT operation, called Doppler-FFT, is applied to resolve multiple objects and calculate the phase difference for each. The velocity of the $i$-th object can then be derived as:

$$v_i = \frac{\lambda \omega_i}{4\pi T_c}, \tag{1.10}$$

where $\lambda$ is the wavelength, $\omega_i$ is the phase difference between consecutive chirps of the $i$-th object, and $T_c$ is the chirp duration.

Velocity estimation provides crucial information about the motion of targets. By estimating velocity, a radar describes how objects are moving, supporting a wide range of motion-based applications. Moreover, it distinguishes multiple objects at

## 1.2 mmWave Sensing Technology

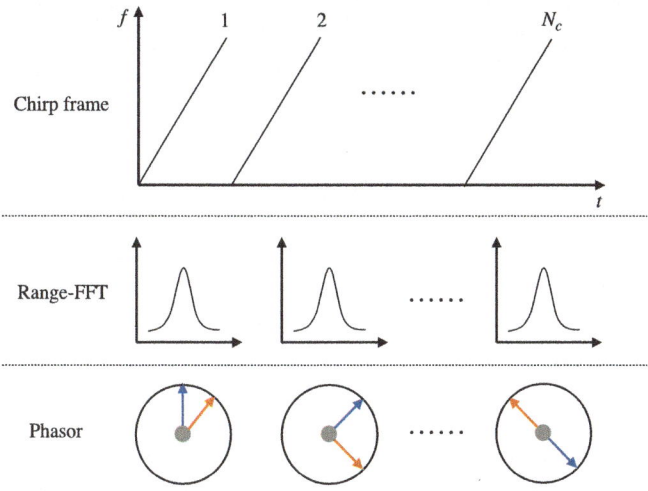

**Fig. 1.3** A chirp frame generates multiple peaks in Range-FFT, each with a different phase, enabling velocity estimation

the same range based on their distinct radial velocities relative to the radar, enabling more precise sensing of multiple targets.

**Angle Estimation** The angle of signals arriving at a radar is also referred to as the angle-of-arrival (AoA). If signals are reflected by an object, the object's angle relative to the radar can be estimated accordingly. Unlike range and velocity estimation using FMCW techniques, angle estimation leverages information from multiple receive antennas. Consider a simple scenario where a mmWave radar is equipped with two receive antennas. The signals reflected by an object arrive at the two antennas along paths of different lengths due to the object's relative angle to the radar, as illustrated in Fig. 1.4. The phase difference between the two paths is given by:

$$\Delta\phi = \frac{2\pi \Delta d}{\lambda}, \qquad (1.11)$$

where $\Delta\phi$ is the phase difference, $\Delta d$ is the difference in path lengths, and $\lambda$ is the wavelength.

Since the distance between the two receive antennas is small compared to the paths' lengths, the paths can be approximated as parallel lines. Consequently, there is a geometrical relationship between the paths' lengths and the angle: $\Delta d = l \sin(\theta)$, where $l$ is the distance between the two antennas, and $\theta$ is the object's angle. Using this relationship, the object's angle can be calculated as:

$$\theta = \sin^{-1}\left(\frac{\lambda \Delta\phi}{2\pi l}\right), \qquad (1.12)$$

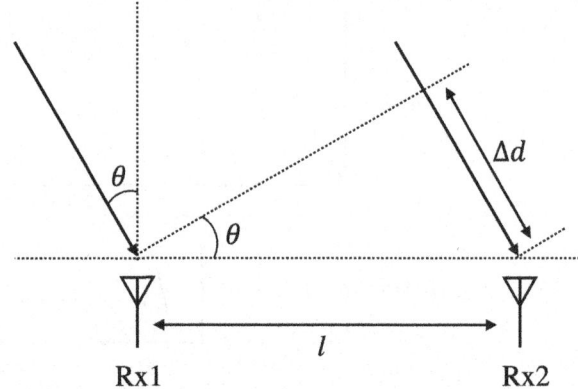

**Fig. 1.4** Signals arrive at the two receive antennas with different path lengths

where $\theta$ is the angle. Based on the properties of the trigonometric function, $\sin(\theta)$ and $\theta$ exhibit similar trends for small values of $\theta$. This implies that angle estimation is more precise when the object is located at a small angle relative to the radar [9].

To estimate the angles of multiple objects, radars require an antenna array consisting of multiple receive antennas and utilize the phase differences across antennas. Suppose there are two objects with different angles relative to a radar equipped with $N_{RX}$ receive antennas. The signals reflected by the objects are superimposed at each antenna, producing $N_{RX}$ phasors. The received signals from the antennas can be treated as a discrete sequence, where each antenna's signal contains two phasors of different angular frequencies. A Fourier transform is then applied to this sequence to extract the angular frequencies of the phasors. Using the extracted frequencies, the angle of the $i$-th object is determined as:

$$\theta_i = \sin^{-1}\left(\frac{\lambda \omega_i}{2\pi l}\right), \qquad (1.13)$$

where $\theta_i$ is the angle of the $i$-th object, $\lambda$ is the wavelength, $\omega_i$ is the angular frequency derived by FFT, and $l$ is the distance between adjacent antennas. This process is called Angle-FFT.

Angle estimation provides an additional spatial perspective for sensing objects. By combining angle and range information, mmWave radars can localize objects and describe their shapes, enabling mmWave radar-based localization and object detection applications.

### 1.2.3 mmWave Sensing Applications

With the continuous development of mmWave sensing technology and the increasing popularity of wireless devices, a great diversity of mmWave sensing applications has emerged in recent years. These applications are mainly involved with three application scenarios, i.e., autonomous driving, smart homes, and industrial manufacturing [10].

**Autonomous Driving** Utilizing radars to sense the environment for automotive driving is one of the original tasks of mmWave radars. In automotive-related scenarios, object detection is a critical task that leverages mmWave radars to detect and sense various vehicles, pedestrians, and roadblocks. It uses point clouds or heatmaps from mmWave signals and subsequently localizes the bounding boxes of target objects. The complementary strengths of mmWave radars and cameras have inspired fusing these modalities for enhanced object detection. Ego-motion estimation focuses on estimating the motion parameters of vehicles or robots using built-in mmWave radars, which is the basis for navigation and interaction with the environment. Simultaneous localization and mapping (SLAM) represents another essential application in automotive scenarios. SLAM emphasizes building a map of the environment and targets in addition to localization. mmWave-based SLAM can support applications in robotics and autonomous systems.

**Smart Homes** The integration of mmWave RF modules with IoT devices in indoor environments has enabled a variety of smart home applications. These applications focus on providing human-centric services, including user monitoring and security surveillance. Activity recognition is a common task, which leverages mmWave signals to sense human activities and classify various human activities into walking, eating, typing, etc. This plays a vital role in understanding human intentions in smart homes. Pose estimation reconstructs a person's pose by estimating the coordinates of their skeleton joints (typically 17, 19, or 25 joints) based on mmWave signals, which is a more detailed and visual presentation of human activities. Gesture recognition uses mmWave radar sensing to classify hand gestures or estimate hand positions. Reflected mmWave signals are processed to identify gesture types or locations. Gesture recognition plays an important role in human-computer interaction. Speech recognition utilizes mmWave signals reflected by the human mouth, throat, or speakers to capture vibration information, enabling speech recognition and content recovery. Vital sign monitoring is a common healthcare-related application. Since mmWave signals have short wavelength, they can detect subtle movement in human chest or skin. Hence, using mmWave can monitor respiration and heartbeat activities. mmWave-based User authentication is an innovative application that provides contactless security. The application leverages mmWave signals to sense human behavior or biometrics, and extracts behavioral or biometric features for classifying user identity. Indoor positioning is a fundamental smart home function, which provides necessary location information of the target users or devices. This is achieved using range and angle information estimated by mmWave radars.

**Industrial Manufacturing** mmWave radars serve as versatile sensors in industrial applications, offering high sensing capability, privacy preservation, and low cost. The characteristics make mmWave radars suitable for replacing traditional sensors in privacy-sensitive scenarios (e.g., imaging) and for widespread use in factories to enable manufacturing assistance. Recent advancements have led to a variety of industrial applications. Industrial imaging uses mmWave radars to derive spatial information, such as azimuth and elevation angles, and generate images of targets. This imaging capability extends usage to scenarios where cameras are less effective, such as environments with barriers or poor weather conditions. Industrial measurement takes advantage of the millimeter-level wavelength and centimeter-level range estimation capabilities of mmWave radars. These radars are used to measure physical parameters with high precision, enabling novel industrial applications. Environmental monitoring leverages mmWave radars as environmental sensors to monitor weather, soil conditions, liquids, or detect insects. For instance, weather monitoring is often achieved through joint radar and communication, exploiting the impact of environmental factors such as rain and clouds on mmWave communication links. These applications support climate prediction and agricultural productivity.

## 1.3 Organization of Monograph

The monograph has 6 chapters. This chapter provides an introduction to mmWave-based human reconstruction. The chapter starts with an overview of human reconstruction, which includes application scenarios and existing technologies. Then, mmWave sensing is introduced and the key technologies including FMCW technique and sensing parameter estimation are given. Emerging application scenarios of mmWave sensing are presented, covering automotive applications, smart homes, and industrial applications.

Chapter 2 introduces mmWave-based human posture reconstruction system, $m^3Track$, which can reconstruct and track multi-user 3D body postures. The system first separates individual users on radar signals. Then, $m^3Track$ extracts each user's shape and motion features and designs the techniques to reconstruct 3D postures. Furthermore, it maps the reconstructed 3D postures into 3D space and utilizes a coordinate-corrected tracking method to accurately track user positions. A thorough evaluation showcases the practicality of the system in multi-user human posture reconstruction.

Chapter 3 introduces a 3D facial reconstruction system, *mm3DFace*, which utilizes an mmWave radar to reconstruct dynamic human faces displaying facial expressions. The system processes the pre-captured mmWave radar signals and extracts key facial geometric features that reflect subtle changes in facial expressions. Following this, *mm3DFace* derives robust facial shapes that are invariant to distance and orientation, leveraging 68 facial landmarks. Subsequently, the system enhances the reconstructed facial expressions by applying a region-based amplification. Finally, *mm3DFace* presents the way to generate 3D facial avatars. The evaluation results show the effectiveness and practicality of the system for real-world 3D facial reconstruction applications.

Chapter 4 introduces *mmHand*, a novel system for 3D hand pose estimation based on mmWave sensing. The system first uses mmWave signals to detect the user hand and processes the acquired sensing data. Then, it applies a custom-designed attention-driven hourglass network, *mmSpaceNet*, to capture spatial features and employs long short-term memory (LSTM) networks to extract temporal characteristics. Then, the system predicts hand joint locations in 3D space for accurate 3D hand skeleton estimation. Finally, the 3D hand meshes are reconstructed through a hand model. Evaluations show the effectiveness of *mmHand* in hand pose estimation and hand mesh reconstruction.

Chapter 5 introduces state-of-art research related to the monograph. This chapter first introduces human reconstruction research. Then, it provides a review on mmWave sensing research, and then elaborates on mmWave-based human reconstruction research.

Chapter 6 provides a conclusion of the monograph summarizing the content and contribution of the monograph. The conclusion also shows that techniques presented in this book identify new problems and solutions of mmWave-based human reconstruction. We present the future research direction at the end of the monograph.

# References

1. Samavati, T., Soryani, M.: Deep learning-based 3d reconstruction: a survey. Artif. Intell. Rev. **56**(9), 9175–9219 (2023)
2. Moniruzzaman, M., Yin, Z., Hossain, M.S.B., Choi, H., Guo, Z.: Wearable motion capture: reconstructing and predicting 3d human poses from wearable sensors. IEEE J. Biomed. Health Inf. **27**(11), 5345–5356 (2023)
3. Yu, Q., Han, C., Bai, L., Choi, J., Shen, X.: Low-complexity multiuser detection in millimeter-wave systems based on opportunistic hybrid beamforming. IEEE Trans. Veh. Technol. **67**(10), 10129–10133 (2018)
4. Qiao, J., Shen, X., Mark, J.W., Shen, Q., He, Y., Lei, L.: Enabling device-to-device communications in millimeter-wave 5g cellular networks. IEEE Commun. Mag. **53**(1), 209–215 (2015)
5. He, S., Zhang, Y., Wang, J., Zhang, J., Ren, J., Zhang, Y., Zhuang, W., Shen, X.: A survey of millimeter-wave communication: physical-layer technology specifications and enabling transmission technologies. Proc. IEEE **109**(10), 1666–1705 (2021)
6. Instruments, T.: mmwave radar sensors (2024). https://www.ti.com/sensors/mmwave-radar/overview.html
7. Meta, A., Hoogeboom, P., Ligthart, L.P.: Signal processing for fmcw sar. IEEE Trans. Geosci. Remote Sensing **45**(11), 3519–3532 (2007)
8. Duhamel, P., Vetterli, M.: Fast fourier transforms: a tutorial review and a state of the art. Signal Process. **19**(4), 259–299 (1990)
9. Klukas, R., Fattouche, M.: Line-of-sight angle of arrival estimation in the outdoor multipath environment. IEEE Trans. Veh. Technol. **47**(1), 342–351 (1998)
10. Kong, H., Huang, C., Yu, J., Shen, X.: A survey of mmwave radar-based sensing in autonomous vehicles, smart homes and industry. IEEE Commun. Surveys Tutor. **27**(1), 463–508 (2024)

# Chapter 2
# mmWave-based Human Posture Reconstruction

**Abstract** 3D human posture reconstruction has rapidly expanded across a wide range of applications. However, current vision-based posture tracking systems face significant challenges, such as privacy concerns and reliance on favorable lighting conditions. To address these issues, recent studies have explored the use of commodity radio frequency signals for 3D human posture tracking, offering a more privacy-preserving and robust solution. Despite these advancements, existing methods struggle to handle scenarios with multiple users in the same environment. In this chapter, we introduce $m^3 Track$, which employs a single mmWave radar to simultaneously reconstruct and track the postures of multiple users as they move, walk, or sit. Leveraging sensing signals from a mmWave radar in multi-user environments, $m^3 Track$ first isolates individual users on the radar signals. It then extracts each user's shape and motion features and reconstructs their 3D postures using a specially designed deep learning model. Afterward, $m^3 Track$ maps the reconstructed 3D postures into 3D space and utilizes a coordinate-corrected tracking method to accurately track user positions. Experiments conducted in real-world multi-user scenarios demonstrate the accuracy and robustness, showcasing its practicality for multi-user 3D posture tracking.

**Keywords** Human posture reconstruction · Millimeter wave · Multi-user scenarios · Deep learning

## 2.1 Introduction

The development of 3D human posture reconstruction technology has advanced rapidly in recent years. The technology generates dynamic and realistic human body skeletons that follow individuals as they move, walk, or sit. With the increasing prevalence of IoT devices, the application scope of 3D posture reconstruction has expanded from specialized use cases such as filmmaking, athletic training, and military applications to broader commercial scenarios, including VR, AR, motion-sensing gaming, and smart-home control. Currently, vision-based approaches dominate the 3D posture reconstruction due to their non-intrusive nature compared to wearable sensors [1, 2]. However, vision-based approaches are highly dependent on environmental lighting conditions and pose significant privacy concerns, which are increasingly scrutinized.

In pursuit of more robust and privacy-preserving methods, researchers have turned to radio frequency signals, such as Wi-Fi [3] and mmWave [4–6], as alternatives for 3D human posture reconstruction. However, these approaches primarily focus on estimating the posture of a single user, leaving multi-user scenarios largely unexplored. Although some studies [7, 8] enable indoor localization in multi-user environments, they are limited to tracking users' locations and cannot generate dynamic skeletons that follow their movements. Other research works [9–11] have demonstrated the potential of multi-user 3D posture reconstruction using RF signals but rely on specialized hardware setups, such as systems with over 20 antennas, making them impractical for widespread deployment.

A robust and practical solution for multi-user 3D posture reconstruction based on COTS RF technology is therefore highly desirable. Such a system would enable easy deployment in real-world scenarios, unlocking applications such as multi-user gaming and collaborative motion tracking. However, achieving multi-user 3D posture reconstruction using a single COTS mmWave radar presents several challenges: (1) Accurately separating multiple users and extracting their individual posture information; (2) Reconstructing fine-grained 3D postures for each user from implicit mmWave signals; (3) Simultaneously tracking the 3D positions of multiple users.

In this chapter, we propose $m^3Track$, a mmWave-based multi-user 3D posture reconstruction and tracking system that leverages a single COTS mmWave radar to simultaneously track the postures of multiple users. $m^3Track$ achieves this through the following key components: (1) User Separation: Utilizing designed chirp signals and minimum variance distortionless response (MVDR) to separate multiple users on mmWave signals; (2) Posture Reconstruction: Extracting spatial and temporal features from separated mmWave signals and employing a forked ConvLSTM deep learning model to reconstruct the 3D posture of each user; (3) 3D Space Mapping and Tracking: Mapping reconstructed postures into real-world 3D space by minimizing mapping errors between the postures and the corresponding point clouds, and tracking user positions using a coordinate-corrected extended Kalman filter.

To evaluate the performance of $m^3Track$, we conducted experiments in six real-world multi-user scenarios using a COTS mmWave radar. The results demonstrate that $m^3Track$ can effectively track up to four users simultaneously, achieving an average joint tracking error of 42.4 mm and a localization error of 21.5 mm.

We summarize our main contributions as follows:

- We design $m^3Track$, a practical multi-user 3D posture reconstruction system using a single COTS mmWave radar, enabling a wide range of real-world applications for multi-user scenarios.
- We propose a novel multi-user separation method for effectively isolating users in mmWave signals and develop a ConvLSTM-based deep learning model to reconstruct 3D postures for each user.

## 2.2 System Overview

- We present a point cloud-based mapping approach to place all reconstructed 3D postures into real-world 3D space and introduce a coordinate-corrected tracking method for continuous multi-user posture tracking.
- We validate $m^3Track$ through extensive experiments, demonstrating its accuracy and robustness in multi-user posture reconstruction in real-world environments.

## 2.2 System Overview

The system framework of $m^3Track$ is shown in Fig. 2.1. In $m^3Track$, a mmWave radar propagates chirp signals into the sensing area, where these signals are reflected by objects. The reflected signals are then captured by the radar's receive antennas, providing the raw data needed to sense all objects within the coverage area. $m^3Track$

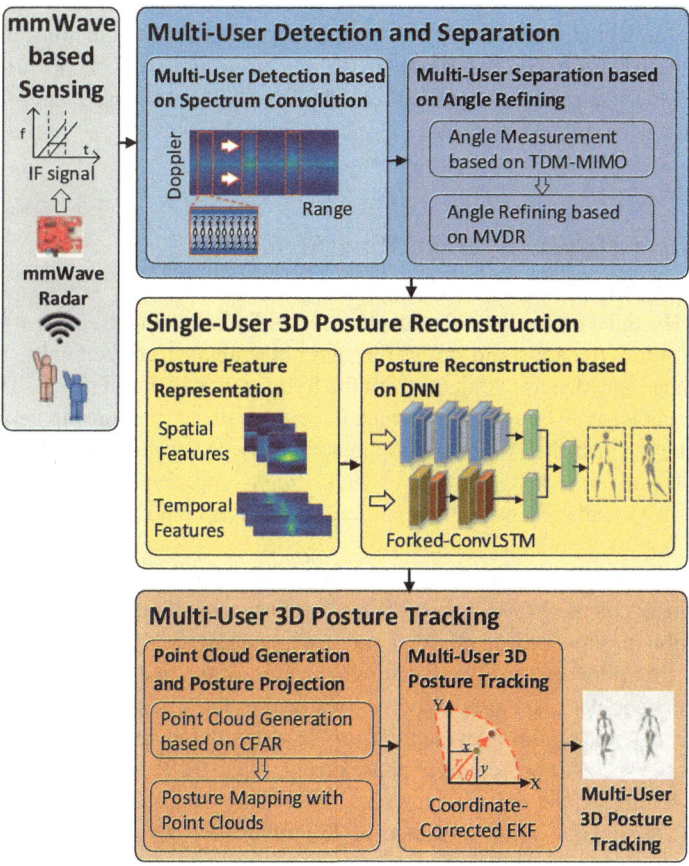

**Fig. 2.1** System framework of $m^3Track$

employs a custom-designed spectrum convolution method to detect all users in the sensing area. Then, a MVDR approach is applied to separate the individual users from the mixed mmWave signals. After separating the mmWave signal profiles for each user, $m^3$Track extracts spatial features that capture a user's shape and temporal features that describe their motion. Using these features, $m^3$Track reconstructs the 3D human posture of each user through a specially designed deep neural network, a forked ConvLSTM model. In multi-user scenarios, $m^3$Track generates point clouds representing all users and aligns the reconstructed 3D postures with these point clouds in 3D space by minimizing the mapping errors between them. To track the positions of multiple users over time, $m^3$Track employs a coordinate-corrected extended Kalman filter, enabling robust and practical 3D posture tracking for all users.

## 2.3 Multi-User Detection and Separation

Using the collected mmWave signals, $m^3$Track captures all dynamic and static objects within the sensing environment. To perform multi-user 3D posture reconstruction, $m^3$Track removes static objects, detects all potential users, and separates multiple users on mmWave signals.

### 2.3.1 User Detection on mmWave Signals

Since the IF signal of mmWave reveals the range of sensed objects, $m^3$Track detects all objects (both dynamic and static) through range analysis. Specifically, $m^3$Track extracts the range of sensed objects from the frequency $f$ of the IF signal. The range $r$ of an object is given by $r = \frac{cfT_c}{2B}$, which indicates a linear relationship between the object's range $r$ and the IF signal's frequency $f$. By performing FFT (Range-FFT) on the IF signal, $m^3$Track generates a *Range-Profile* representing the objects sensed by the mmWave radar. Thus, $m^3$Track initially detects all objects in the sensing area, including both dynamic and static ones.

To enable human posture reconstruction, $m^3$Track removes static objects and identifies users on mmWave signals. Typically, background objects are static, while users exhibit motion—even minimal movements such as breathing or heartbeat. Based on this principle, $m^3$Track distinguishes users by analyzing object motion. It leverages Doppler responses, which correspond to movement speeds, to filter out static objects while retaining dynamic users. To extract Doppler responses, $m^3$Track calculates the speed of objects from phase changes $\omega$ in the IF signal. The speed $v$ of an object relative to the radar is given by $v = \frac{\lambda}{4\pi T_c}\omega$, where $\lambda$ is the wavelength of the mmWave signal. Similar to Range-FFT, another FFT (Doppler-FFT) is applied to the *Range-Profile*, resulting in a *Range-Doppler-Profile* that captures the speed of all objects at different ranges.

2.3 Multi-User Detection and Separation

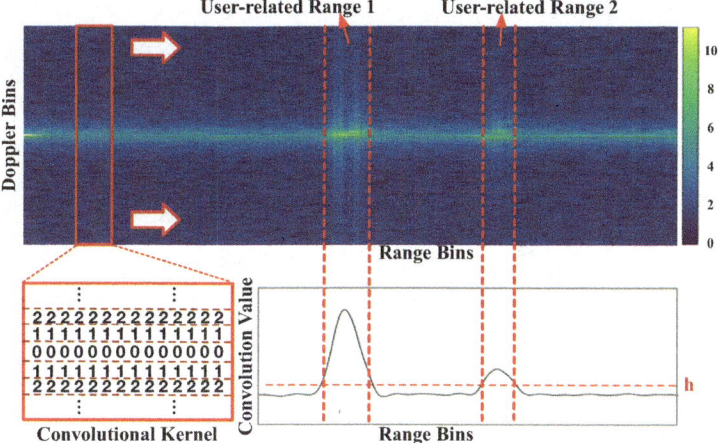

**Fig. 2.2** Illustration of user detection

Using the range and speed data, $m^3Track$ distinguishes users from background objects via a convolution-based approach, referred to as spectrum convolution, as illustrated in Fig. 2.2. A specialized convolutional kernel slides across the range bins of the *Range-Doppler-Profile*, performing a convolution operation at each step to detect users. The kernel has a specific width equal to the *Range-Doppler-Profile* width and a length of 14, corresponding to a 0.5 m range resolution. This resolution slightly exceeds the typical width of a human body, enabling the kernel to cover individual users on mmWave signals. The kernel's design assigns weights to Doppler responses, emphasizing dynamic objects while de-emphasizing static ones. Rows closer to the kernel center are assigned lower weights, while rows farther from the center are assigned higher weights. This configuration ensures that $m^3Track$ focuses on ranges with significant Doppler responses, capturing users' motions while ignoring static objects.

After performing spectrum convolution, $m^3Track$ computes convolution values for different ranges and uses an empirical threshold $h$ to identify user ranges. Figure 2.2 demonstrates a scenario where two users are detected at different ranges. Even if a user remains stationary (e.g., user 2), their minor movements from breathing or heartbeat produce higher convolution values than static background objects.

### 2.3.2 User Separation on mmWave Signals

In real-world scenarios, multiple users may occupy the same range bin and lead to ambiguities in range dimension of mmWave signals, as depicted in Fig. 2.3a. However, users sharing the same range exhibit distinct angular positions relative to

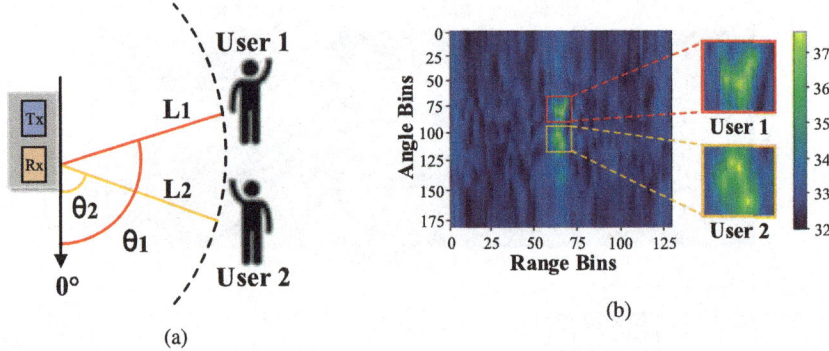

**Fig. 2.3** Illustration of user separation. (**a**) Users of the same range. (**b**) MVDR-based *Range-Angle-Profile* viewed from elevation

the radar, providing a solution to separate these users on mmWave signals. Therefore, to isolate the mmWave profiles corresponding to individual users in multi-user scenarios, we leverage their angular differences as an additional discriminative feature.

The utilization of multiple transmit and receive antennas in mmWave radar facilitates the estimation of angles. The orientation of an object relative to the mmWave radar, described by its azimuth angle $\theta$ and elevation angle $\phi$, can be calculated as follows:

$$\theta = \sin^{-1}\left(\frac{\lambda \omega_a}{2\pi d_1}\right), \quad \phi = \sin^{-1}\left(\frac{\lambda \omega_e}{2\pi d_2}\right), \tag{2.1}$$

where $d_1$ and $d_2$ represent the physical separations of the receiving antennas along the azimuth and elevation axes, respectively. The terms $\omega_a$ and $\omega_e$ correspond to the phase differences across multiple-input and multiple-output (MIMO) channels in the azimuth and elevation directions. Through AoA estimation, we obtain *Angle-Profile*, which provides angular information for all detected objects. By combining the *Range-Profile* with the *Angle-Profile*, a *Range-Angle-Profile* is constructed, enabling the differentiation of individual users.

To compute the angles of multiple users, $m^3Track$ employs the TDM-MIMO approach using a mmWave radar. Specifically, a COTS mmWave radar (TI IWR1443BOOST), features 3 transmit antennas (Tx) and 4 receive antennas (Rx). These antennas form a 3 × 4 transmit-receive pair configuration, which can be expanded into a 2D multiple-input and multiple-output (MIMO) array, as illustrated in Fig. 2.4. This arrangement allows $m^3Track$ to estimate target angles in both azimuth and elevation planes.

In $m^3Track$, the 2D MIMO array operates under a time-division multiplexing (TDM) scheme, where each transmit antenna (Tx1, Tx2, and Tx3) is sequentially activated in alternating time slots, while all 4 receive antennas simultaneously capture the reflected mmWave signals. By leveraging the TDM-MIMO mechanism,

## 2.3 Multi-User Detection and Separation

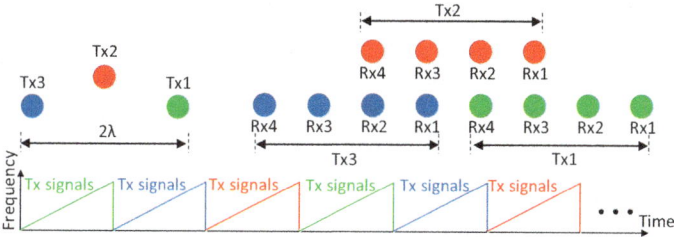

**Fig. 2.4** Illustration of antenna array for $m^3 Track$

$m^3 Track$ mitigates the hardware limitation of a limited number of physical antennas, creating virtual antenna arrays. These virtual arrays provide the foundation for accurately extracting angular information of multiple users from mmWave signals.

To enhance the precision of angle estimation for different users, we apply the MVDR method [12]. The core principle of MVDR is to suppress interference and noise originating from undesired angles while maintaining a distortionless response toward the target angle. By integrating the angles derived from the TDM-MIMO method, $m^3 Track$ achieves a more refined angle extraction. In MVDR, let $a_\theta$ denote the steering vector corresponding to the angle of arrival $\theta$. The MVDR weight vector is computed as:

$$w = \frac{R^{-1} a_{\theta,\phi}}{a_{\theta,\phi}^H R^{-1} a_{\theta,\phi}}, \qquad (2.2)$$

where $R$ represents the correlation matrix of the antenna array. Using the optimized weight vector from MVDR, the signal power output of the antenna array is expressed as:

$$P_{MVDR}(\theta, \phi) = \frac{1}{a_{\theta,\phi}^H R^{-1} a_{\theta,\phi}}, \qquad (2.3)$$

which is determined by identifying peaks in the angular spectrum.

By replacing the conventional *Angle-Profile* with the signal power calculated using Eq. (2.3) for each range bin, $m^3 Track$ achieves fine-grained angle resolution. Figure 2.3b illustrates the MVDR-based *Range-Angle-Profile* for two users located at the same range viewed from an elevation perspective. Although the users share the same range, they are distinctly separated based on their angles in the MVDR-enhanced *Range-Angle-Profile*. This demonstrates the effectiveness of the MVDR approach in resolving angular ambiguities and separating multiple users in mmWave scenarios. By utilizing both range and angle information, $m^3 Track$ isolates and extracts the mmWave profiles corresponding to individual users in multi-user environments. These profiles are subsequently employed to enable 3D posture reconstruction for multiple users simultaneously.

## 2.4 Single-User Posture Reconstruction

Once individual users are separated on mmWave signals in multi-user scenarios, $m^3$ Track further reconstructs the 3D posture of each user. This step is essential for achieving accurate 3D posture tracking of multiple users.

### 2.4.1 Posture Feature Representation

To reconstruct 3D human posture, $m^3$ Track extracts spatial features that capture the user's shape and temporal features that represent the user's motion from the mmWave signals.

To detect and differentiate multiple users, $m^3$ Track determines the range and angles (azimuth and elevation) of each user relative to the radar. Using this information, $m^3$ Track generates feature patterns for individual users within the *Range-Angle-Profiles*. For a user $i$, let the range be $r_i$, the azimuth angle $\theta_i$, and the elevation angle $\phi_i$. The central coordinate for the user can then be represented as $(r_i, \theta_i, \phi_i)$. To model the user's shape, the central coordinate is used as an anchor, around which multiple cylinders are constructed to encompass the human body, as illustrated in Fig. 2.5a. Specifically, a 3-cylinder model $M(C_h, C_t, C_l)$ is employed: the head cylinder $C_h(r_i, \theta_i, \phi_i)$ corresponds to the head and neck region, the torso cylinder $C_t(r_i, \theta_i, \phi_i)$ represents the chest and arms, and the leg cylinder $C_l(r_i, \theta_i, \phi_i)$ covers the lap and legs.

The radius and height of these cylinders adapt dynamically to fit the corresponding body regions. The torso cylinder's height is fixed around the anchor point since the torso's vertical length exhibits minimal variation across different postures and individuals. The other two cylinders adjust proportionally to cover the remaining body regions. For instance, with the torso cylinder height set to 60 cm, the head and leg cylinders expand vertically to the margins of the feature pattern. Each cylinder is localized in the human body model, and for each one, $m^3$ Track computes two *Range-Angle-Profiles* based on azimuth and elevation angles, respectively, as shown in Fig. 2.5a. These profiles are later combined within a neural network model, making minor deviations in cylinder proportions from actual body regions inconsequential. In total, 6 *Range-Angle-Profiles* (two per cylinder) are extracted for each user and serve as spatial features. This approach of calculating profiles for individual body parts enhances the representation of specific regions, offering a detailed depiction of the user's shape in 3D space. These spatial features form the foundation for 3D posture reconstruction.

In addition to spatial features, temporal features play a crucial role in 3D posture reconstruction, as they capture the dynamic movement of various body parts over time for each individual. To extract temporal features, $m^3$ Track employs a 3-cylinder model. For every cylinder, a *Range-Doppler-Profile* is computed to represent the motion of the corresponding body parts. Specifically, for two consecutive time slots,

## 2.4 Single-User Posture Reconstruction

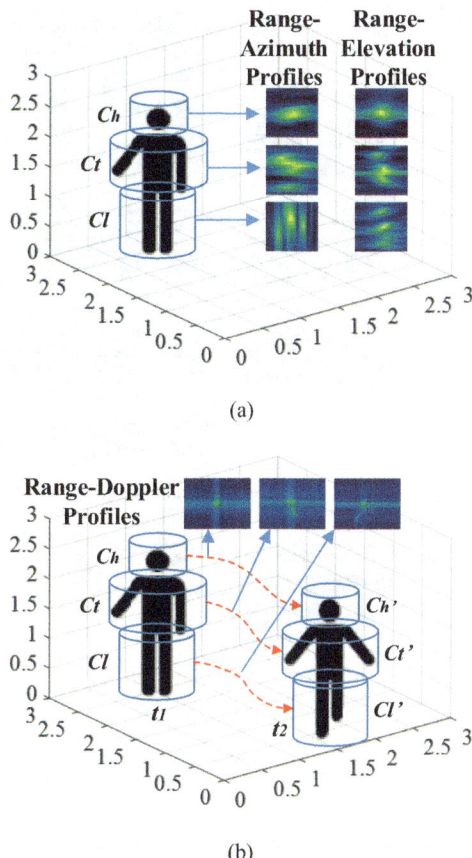

**Fig. 2.5** Illustration of posture feature representation for spatial and temporal features. (**a**) Spatial features. (**b**) Temporal features

$t_1$ and $t_2$, the *Range-Doppler-Profiles* illustrate the movement of distinct body parts as they evolve over time, as demonstrated in Fig. 2.5b. In total, 3 distinct *Range-Doppler-Profiles* are derived from the mmWave signals for each user, capturing the temporal motion of various body parts. These temporal features are then used as a fundamental input for 3D posture reconstruction.

As a result, the spatial features capture the shape of each user, while the temporal features describe their motion patterns in multi-user environments. These features are then utilized for 3D posture reconstruction.

### 2.4.2 3D Posture Reconstruction

We propose a deep learning model, specifically a forked-ConvLSTM, which maps spatial and temporal features to the user's skeleton joint coordinates for 3D posture reconstruction. The architecture of the proposed model is shown in Fig. 2.6. The

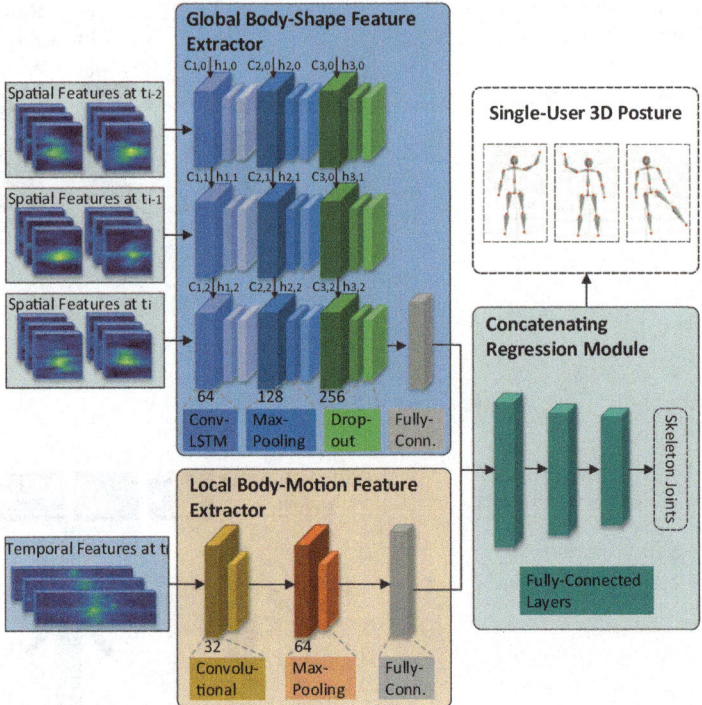

**Fig. 2.6** Neural network structure of $m^3Track$

model processes the sequence of spatial features and their corresponding temporal features through two distinct branches. To extract feature embeddings from the spatial and temporal features, we design a *Global Body-Shape Feature Extractor* and a *Local Body-Motion Feature Extractor*, respectively. These extracted embeddings are then combined in a *Concatenating Regression Module*, which predicts the 3D coordinates of the user's skeleton joints, enabling the reconstruction of 3D posture. Since 3D posture reconstruction is formulated as a regression task, the model can handle various postures to ensure versatility.

The *Global Body-Shape Feature Extractor* is designed to capture the body-shape feature embeddings from the spatial features of a user. At any given moment, the spatial features, i.e., the *Range-Angle-Profiles*, reflect the global shape of the target user's body in space, representing the instantaneous configuration of their body shape. In the subsequent moment, the *Range-Angle-Profiles* evolve from the previous state, indicating that each part of the profile is dependent on the preceding moment. This reveals that the spatial features are interconnected both in time and space, motivating the use of convolutional operations to learn spatial regional characteristics, alongside memory cells to model temporal dependencies. Based on this insight, we adopt ConvLSTM [13], a combination of convolutional neural

## 2.4 Single-User Posture Reconstruction

networks (CNNs) and LSTM, as the foundation for extracting the global body-shape features.

The architecture of the *Global Body-Shape Feature Extractor* is depicted in Fig. 2.6, which comprises three layers of ConvLSTM. The input to the model consists of multiple profiles, which are concatenated from the posture feature representation, and passed into the input layer. To capture the temporal dependencies, the spatial features at times $t_{i-2}$, $t_{i-1}$, and $t_i$ are simultaneously fed into the network. The output from the final node of the third layer is then taken as the feature extraction result for time $t_i$. Additionally, a convolutional operation is employed to assign regional attention to each spatial feature, while preserving the temporal relationships. The output can be expressed as:

$$Z' = \text{FC}\left(\text{CL}\left(F_m(t), F_m(t-1), \cdots, \Theta'\right)\right), \tag{2.4}$$

where $Z'$ represents the output feature embeddings, $\text{FC}(\cdot)$ denotes the fully-connected layer, $\text{CL}(\cdot)$ stands for the convolutional long short-term memory network operation, and $\Theta'$ are the trainable parameters. At the end of the feature extraction process, a fully connected network stretches the feature embeddings to facilitate the subsequent feature concatenation operation. Finally, the extracted feature embeddings, which describe the global body shape of the user, are passed to the following Concatenating Regression Module for further processing.

The *Local Body-Motion Feature Extractor* is designed to capture the feature embeddings corresponding to the local body motions of a user, based on the temporal features. The temporal features, i.e., the *Range-Doppler-Profiles*, represent a sequence of local body movements, indicating the velocities of various body parts. Thus, the *Local Body-Motion Feature Extractor* is built to extract feature embeddings related to Doppler responses, which describe the movements of different body parts.

As illustrated in Fig. 2.6, this module consists of a two-layer convolutional neural network that processes the *Range-Doppler-Profiles* $F_d(t)$ of a user to obtain local body-motion feature embeddings. The profiles, derived from the posture feature representation, are concatenated before being passed into the input layer. The output of this module represents the feature embedding of a user's local body movements during the given time period, which is expressed as:

$$Z = \text{Conv}\left(F_d(t), \Theta\right), \tag{2.5}$$

where $Z$ denotes the output feature embedding, $\text{Conv}(\cdot)$ refers to the convolution operation applied to the temporal features, and $\Theta$ represents the trainable parameters. Through the feature extraction in this module, $m^3 Track$ captures the feature embeddings within the temporal features that describe the local body motions of a user. These embeddings are then forwarded to the subsequent Concatenating Regression Module.

In the *Concatenating Regression Module*, the feature embeddings from the previous two modules are merged to form a stitching vector, which is then used to predict the user's skeleton joint coordinates. This module consists of three fully-connected layers, as depicted in the right section of Fig. 2.6. The first layer is responsible for encoding a stitching vector that integrates the feature embeddings from both previous modules. Using this stitching vector, the second and third layers predict the 3D coordinates of the skeleton joints in the Cartesian coordinate system. The operation of the Concatenating Regression Module can be represented as:

$$\hat{P} = G\left(\text{Concat}\left(Z, Z'\right), \hat{\Theta}\right), \tag{2.6}$$

where $\hat{P}$ is the predicted 3D skeleton joint coordinates, $G(\cdot)$ denotes the regression network operation, $\text{Concat}(\cdot)$ refers to the concatenation of features, and $\hat{\Theta}$ represents the trainable parameters. Once the entire deep learning model is trained, the regression module outputs the predicted 3D coordinates for the user's skeleton joints.

To train the deep learning model for predicting the 3D skeleton joints of a user, we define a loss function based on the distance between the predicted skeleton joints and the ground truth coordinates obtained from Kinect. The loss function is formulated using the Smooth $L_1$ loss [14], as follows:

$$L_p = \begin{cases} \frac{1}{2}(P - \hat{P})^2 & \text{if } |P - \hat{P}| \leq \delta \\ \delta(|P - \hat{P}| - \frac{1}{2}\delta) & \text{if } |P - \hat{P}| > \delta \end{cases} \tag{2.7}$$

where $P$ denotes the ground truth 3D skeleton joint coordinates, and $\hat{P}$ represents the predicted 3D skeleton joint coordinates. An outlier threshold $\delta$ is incorporated into the loss function to handle joint detection outliers, preventing the model from non-convergence due to the high penalty associated with outliers. The ground truth 3D skeleton joint coordinates are adjusted by subtracting the center of the user's joint coordinates, yielding the relative positions of the joints. This step helps train a more accurate and robust model, as it removes the influence of the user's absolute position. By optimizing the parameters $\Theta$, $\Theta'$, and $\hat{\Theta}$ with this loss function, the model ultimately predicts the 3D coordinates for each user's skeleton joints in multi-user scenarios.

## 2.5 Multi-User 3D Posture Tracking

While $m^3Track$ reconstructs the 3D postures for each user in multi-user scenarios, these reconstructed postures do not directly correspond to the real-world 3D space. In order to achieve accurate 3D posture tracking, $m^3Track$ must map the reconstructed postures of all users into the real-world 3D space and track the position of each user within this space for effective multi-user 3D posture tracking.

## 2.5.1 Posture Mapping with Point Cloud

To map the postures of all users into the 3D space, $m^3 Track$ first generates point clouds to capture the positional information of all users. It then uses these point clouds to map the reconstructed 3D postures of users into the 3D space.

To generate mmWave-based point clouds, $m^3 Track$ utilizes the constant false alarm rate (CFAR) algorithm [15] to generate point clouds for all users. Once the point clouds for all users are generated, $m^3 Track$ proceeds to map each user's reconstructed posture into the corresponding point clouds in 3D space. However, the mapping relationship between the postures and point clouds in 3D space is initially unknown. To establish this relationship for posture tracking, we aim to minimize the mapping errors between the postures and the point clouds, as illustrated in Fig. 2.7. In particular, for each point cloud, $m^3 Track$ clusters the points into several groups using the K-Means algorithm [16]. Each cluster's center corresponds to a skeleton joint. If the distance between the joints and their respective cluster centers is minimized, $m^3 Track$ considers the posture to be accurately mapped with the point cloud. Thus, the 3D postures of all users are mapped to the point clouds by minimizing the mapping errors between the reconstructed postures and the point clouds. This can be formalized as:

$$m(i,k) = \arg\min \sum_{i=1}^{P} \sum_{j=1}^{J} \left(Joint_{k,j} + \hat{s} - X_{i,j}\right)^2, \quad (2.8)$$

where $m(i, k)$ represents the optimal mapping relation between the point clouds and reconstructed postures, $P$ is the total number of observable point clouds, $J$ is the total number of observable joints, $Joint_{k,j}$ is the $j$-th joint of the $k$-th posture, $\hat{s}$ is the correction displacement, and $X_{i,j}$ is the $j$-th cluster center for the $i$-th point

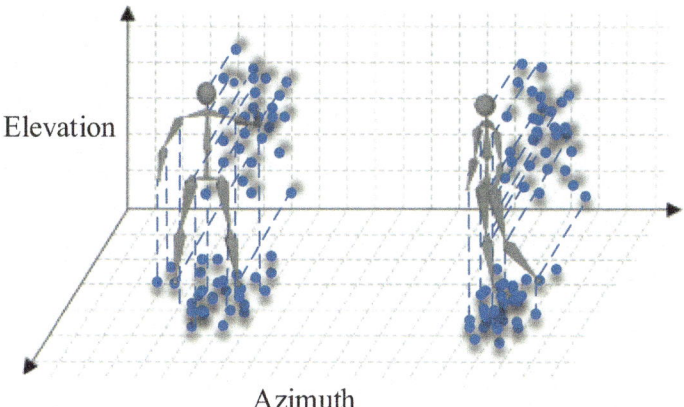

**Fig. 2.7** Multi-user 3D posture mapping

cloud. By solving this optimization problem, we can determine the optimal mapping relationship between all postures and point clouds, allowing the reconstructed 3D postures of all users to be accurately mapped into real-world 3D space.

### 2.5.2 3D Posture Tracking

Using the mapped 3D postures of all users, $m^3 Track$ employs a proposed coordinate-corrected extended Kalman filter (EKF) to track the users' positions in 3D space, thereby enabling multi-user 3D posture tracking. EKF [17] is a well-known state estimation technique used for tracking target positions. The fundamental concept behind EKF is to combine the estimated and measured positions to compute a more accurate target position, thereby reducing the impact of measurement errors and interferences during tracking. However, since mmWave radar operates in the polar coordinate system, while 3D human postures are reconstructed in the Cartesian coordinate system, we introduce a coordinate-corrected extended Kalman filter. This modified EKF applies to both coordinate systems, enabling precise tracking of users' positions in the Cartesian coordinate system.

In the coordinate-corrected EKF, we first define a state function that characterizes the current state of the tracking target. This function can be expressed as:

$$s_t = (x \quad y \quad v \quad \theta \quad \omega), \tag{2.9}$$

where $(x, y)$ represents the target's position in the Cartesian coordinate system, $v$ is the Doppler (or radial speed) in the polar coordinate system, $\theta$ is the deflection angle in the polar coordinate system, and $\omega$ is the deflection angle speed in the polar coordinate system. The geometric relationships of the target's movement in the two coordinate systems are depicted in Fig. 2.8. Using this initial state definition along

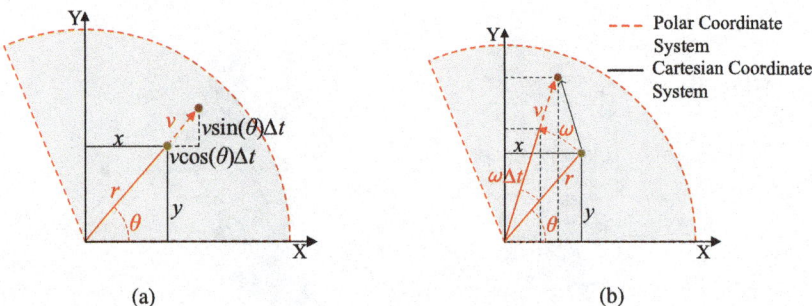

**Fig. 2.8** Relations in two coordinate systems. (**a**) Deflection angle speed $\omega = 0$. (**b**) Deflection angle speed $\omega \neq 0$

## 2.5 Multi-User 3D Posture Tracking

with the geometric relationships, the state function is estimated at each time step $\Delta t$ as follows:

$$s_{t+\Delta t} = \begin{cases} \begin{pmatrix} v\cos(\theta)\Delta t + x \\ v\sin(\theta)\Delta t + y \\ v \\ \theta \\ \omega \end{pmatrix} & \text{if } \omega = 0, \\ \begin{pmatrix} \left(\frac{x}{\cos(\theta)} + v\Delta t\right) \cdot \cos(\omega\Delta t + \theta) \\ \left(\frac{y}{\sin(\theta)} + v\Delta t\right) \cdot \sin(\omega\Delta t + \theta) \\ v \\ \omega\Delta t + \theta \\ \omega \end{pmatrix} & \text{if } \omega \neq 0. \end{cases} \quad (2.10)$$

After the state function is estimated at time $t+\Delta t$, $m^3$Track calculates the covariance of the state function. Subsequently, $m^3$Track measures the ground-truth value of the state function at $t + \Delta t$, denoted as $z_{t+\Delta t}$, and computes its corresponding covariance. Using these values, $m^3$Track combines the estimation and measurement to derive an accurate target state as:

$$\hat{s}_{t+\Delta t} = K_{t+\Delta t} \cdot z_{t+\Delta t} + (I - K_{t+\Delta t}H)s_{t+\Delta t}, \quad (2.11)$$

where $K_{t+\Delta t}$ is the Kalman gain computed from the two covariances, $H$ is a transformation matrix, and $I$ is the identity matrix. By combining the estimation and measurement, $m^3$Track obtains a more precise state (i.e., position) of the target in $\hat{s}_{t+\Delta t}$. Through the iterative application of this process, the position trajectories of user objects are tracked continuously.

By tracking the position trajectories, $m^3$Track ensures continuous monitoring of the mapped postures in the real-world 3D space. However, in practical settings, users might occasionally block each other, resulting in signal propagation blockage, which may cause the mmWave signals to fail in detecting the occluded user. To address this issue and estimate the position of a temporarily blocked user, $m^3$Track relies on the coordinate-corrected EKF's estimation step, specifically using the $(x, y)$ position from $s_{t+\Delta t}$. As a result, $m^3$Track is capable of tracking the position trajectory of each user even during moments of transient blockage. Since the occlusion periods are typically short during movement, and the posture of the blocked user is unavailable, $m^3$Track can still effectively estimate and track the user's position. This approach ensures that robust and practical 3D posture tracking is maintained in multi-user scenarios.

## 2.6 Evaluation

This section presents experimental evaluations to assess the real-world performance of $m^3 Track$.

### 2.6.1 Evaluation Setup

The implementation of $m^3 Track$ utilizes a single COTS mmWave radar, specifically the TI AWR1443 mmWave radar [18], serving as the sensing front-end. This radar features three transmit antennas and four receive antennas, configured to emit mmWave chirp signals with a bandwidth ranging from 77 to 81 GHz. It operates with a signal frame of 128 pulses spanning 50 ms, and each pulse is sampled at a rate of 512 points. To enable high-speed data transfer between the radar and the back-end system, the mmWave radar interfaces with a TI DCA1000EVM data capture card [19]. The back-end processing is conducted on a DELL G15 laptop, which handles the acquisition and analysis of the mmWave data. For ground truth acquisition of human postures, we employ two Kinect V2 devices [20]. These devices utilize RGB and infrared cameras to capture depth images and generate 3D skeleton joints of the users.

The experiments were carried out across various environments, including indoor locations such as a laboratory, corridor, and meeting room, as well as outdoor areas. The environmental setup in the laboratory is illustrated in Fig. 2.9. In detail, an mmWave radar was positioned within the environment, transmitting FMCW signals continuously in the direction of its antenna panel and receiving the reflected signals from users. To record ground truth postures, two Kinect devices were strategically placed at opposite positions: one close to the mmWave radar, and the other 8 m away on the opposite side. Both the mmWave radar and Kinect devices are aligned horizontally at a height of 1.3 m from the ground. Users performed activities within the radar's detection range, at distances varying between 1.2 and 7 m from the radar.

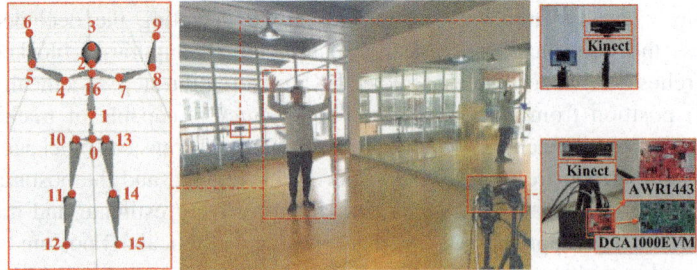

**Fig. 2.9** Experimental settings in the lab

## 2.6 Evaluation

The device placement in all other experimental environments followed the same configuration as that in the laboratory.

We use 17 key skeleton joints of the human body, as depicted in Fig. 2.9. By leveraging these critical joint nodes, which provide structural support for the human body, $m^3Track$ can accurately reconstruct user postures as they engage in activities such as moving, walking, or sitting. A total of 15 volunteers participated in the experiments. The participants performed a variety of natural, everyday activities within the sensing area. These included stationary actions (e.g., arm-lifting, leg-lifting, squatting) and dynamic actions (e.g., walking forward, walking backward, and crossing paths). To ensure consistency, the ground truth posture joints captured by Kinect devices were calibrated by normalizing joint coordinates. This was achieved by subtracting the coordinate center of a user's joints to compute relative positions.

To train the deep learning model, data from two users are utilized. These users perform both stationary activities and walking actions in the laboratory to generate the training dataset. The mmWave data for the two users are collected separately, with each user's dataset comprising 7200 frames captured during the data collection process. The collected mmWave data, along with the skeleton joint coordinates from Kinect as ground truth, are used as inputs for training the deep learning model. The model is trained with a learning rate of 0.001, a batch size of 32, and for 200 epochs. The implementation leverages the Keras framework and is executed on a backend laptop equipped with an Intel i7-11800H processor and NVIDIA RTX3060 GPU. For performance evaluation, data from 13 additional users are employed. These evaluation sessions span approximately 27 hours over 19 days, conducted in diverse environments, including the laboratory, corridor, meeting room, and outdoor areas. The users perform a variety of everyday activities, including stationary and walking tasks, which differ from the specific actions performed during the training phase. To assess the system's capability for multi-user 3D posture tracking, the evaluation scenarios involve groups of users with varying numbers (ranging from 1 to 4) in each session.

### 2.6.2 Overall Performance

We first evaluate the overall performance of $m^3Track$ multi-user 3D posture reconstruction. We provide a visual demonstration of the reconstructed postures in 3D space. As illustrated in Fig. 2.10, the 3D postures for in-place activities are presented alongside the corresponding ground truth and video frames taken within the laboratory environment. The result shows that $m^3Track$ performs well in accurately reconstructing and tracking the 3D postures of individual users. Furthermore, as illustrated in Fig. 2.10b, c, and d, when multiple users engage within the sensing area, $m^3Track$ demonstrates its ability to accurately model the skeletons of each individual. Hence, despite the presence of multiple users, the system efficiently tracks the dynamic postures with minimal interference. These

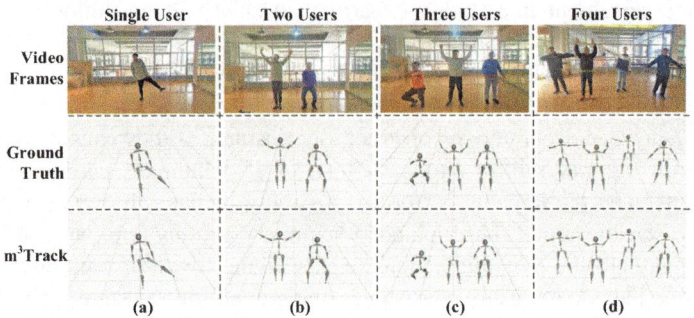

**Fig. 2.10** Multi-user 3D posture tracking for in-place activities

examples show that $m^3Track$ is effective in reconstructing and tracking the 3D postures of multiple users.

In addition to tracking postures for in-place activities, we also evaluate the performance of $m^3Track$ for tracking users while walking. Figure 2.11 illustrates consecutive 3D human postures reconstructed by $m^3Track$ during the tracking of three walking users. It can be seen that the 3D postures of the walking users is reconstructed, capturing intricate details such as arm swings and leg movements. Furthermore, the reconstructed postures consistently align with the real-time positions of the users, demonstrating a continuous tracking of their movements. The results confirm the effectiveness of the proposed solution in mapping the reconstructed postures accurately into real-world 3D space, while continuously tracking the walking trajectories of each individual. As a result, $m^3Track$ is effective in practical multi-user 3D posture reconstruction and tracking.

### 2.6.3 Quantitative Results

To quantitatively evaluate the performance of $m^3Track$, we measure the joint tracking errors, which represent the Euclidean distance between the tracked and ground truth skeleton joint coordinates. These errors are assessed in terms of depth (X), azimuth (Y), and elevation (Z). Table 2.1 presents the joint tracking errors for various user scenarios. The overall joint tracking errors are as follows: 32.4 mm for the 1-user scenario, 34.9 mm for the 2-user scenario, 38.6 mm for the 3-user scenario, and 42.4 mm for the 4-user scenario. These results indicate that $m^3Track$ has small estimation errors in multi-user 3D posture reconstruction and tracking. Also, the increase in tracking errors when multiple users are tracked simultaneously is insignificant compared to single-user scenarios. Furthermore, the errors in depth, azimuth, and elevation exhibit a similar pattern across different numbers of users. These findings highlight that $m^3Track$ is capable of extending 3D posture tracking

2.6 Evaluation

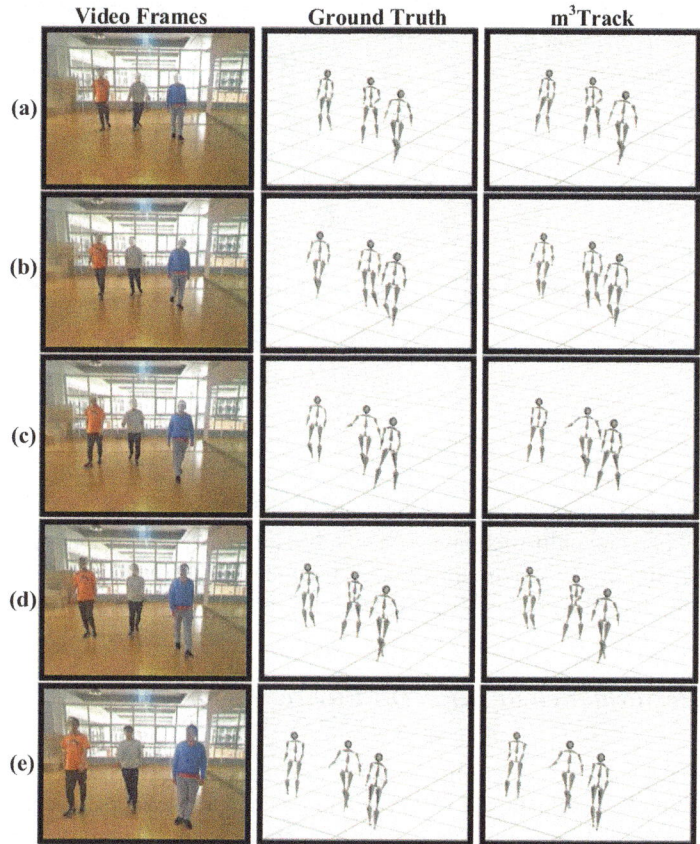

**Fig. 2.11** Multi-user 3D posture tracking for walking activities. (**a**)–(**e**) represent consecutive snapshots during the users' walking process

**Table 2.1** Tracking errors under different user numbers

| Users | 1 | 2 | 3 | 4 |
|---|---|---|---|---|
| Overall | 32.4 mm | 34.9 mm | 38.6 mm | 42.4 mm |
| Depth | 14.4 mm | 15.5 mm | 16.9 mm | 17.4 mm |
| Azimuth | 22.2 mm | 24.6 mm | 28.4 mm | 32.5 mm |
| Elevation | 18.7 mm | 19.3 mm | 20.0 mm | 20.9 mm |

from single-user scenarios to multi-user settings without significant degradation in performance.

Additionally, we delve into the tracking errors of each joint. Figure 2.12 illustrates the joint tracking errors under different user counts. It is evident that the tracking errors differ across various joints. In particular, the joints associated with the arms (Numbers 5, 6, 8, 9) and the legs (Numbers 11, 12, 14, 15) exhibit higher tracking errors, and also experience a more significant increase from single-user to multi-user settings, in contrast to other joints. This is attributed to the more

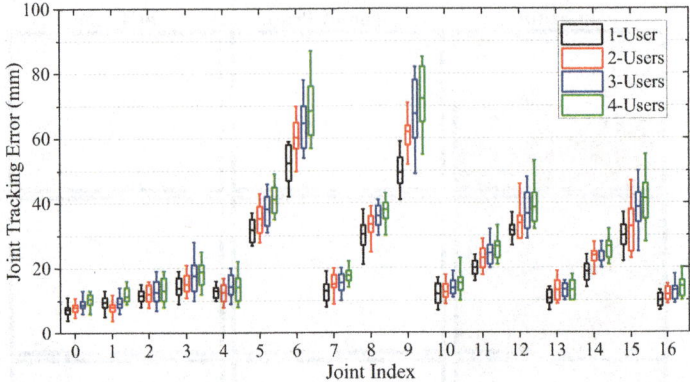

**Fig. 2.12** Joint tracking errors

intricate and variable movements of the limbs, which can sometimes be occluded by other body parts in multi-user environments. Nevertheless, despite the higher errors in tracking limbs, the maximum error in tracking hands remains below 90 mm, highlighting the efficacy of $m^3Track$ in tracking various body parts.

### 2.6.4 Performance in Different Environments

In order to evaluate the versatility and reliability of $m^3Track$ across diverse environments, we conduct evaluation in various environmental settings. These experimental conditions span a laboratory, two corridors, a conference room, and two outdoor areas, each characterized by distinct spatial dimensions and environmental config-

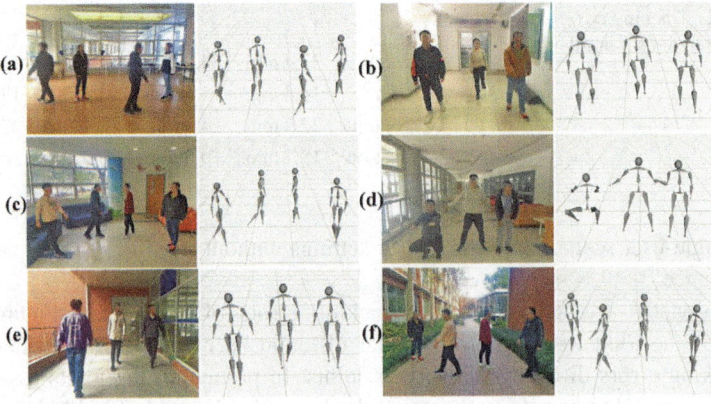

**Fig. 2.13** Multi-user 3D posture tracking in different environments, where (**a**) is in a laboratory, (**b**) is in a corridor, (**c**) is in a conference room, (**d**) is in a corridor, (**e**) and (**f**) are in outdoor areas

## 2.6 Evaluation

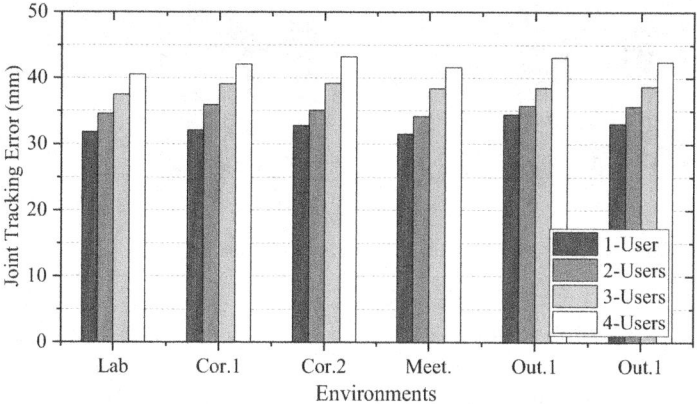

**Fig. 2.14** Joint tracking errors in different environments

urations. Figure 2.13 presents examples of multi-user 3D posture tracking across various environments. In each scenario, $m^3Track$ effectively generates the 3D posture of each user, demonstrating minimal sensitivity to the environmental setup. For example, in Fig. 2.13b, despite the presence of walls adjacent to the users, $m^3Track$ accurately tracks their postures, unaffected by the multipath reflections from the walls. Additionally, in the two outdoor environments depicted in Fig. 2.13e and f, the system precisely tracks the posture of all users, showcasing the effectiveness of multi-user 3D posture reconstruction and tracking even in open spaces.

Next, we perform a quantitative analysis of the multi-user 3D posture tracking performance across various environments. Figure 2.14 illustrates the joint tracking errors for different user scenarios in the six environments. It can be seen that the joint tracking errors remain consistent across the environments with small variation. In particular, the standard deviation of joint tracking errors across these environments is only 3.2 mm. Therefore, $m^3Track$ demonstrates effective and reliable multi-user 3D posture reconstruction and tracking in different environmental settings.

### 2.6.5 Performance in Occluded Scenarios

To evaluate the performance of $m^3Track$ under NLOS conditions, we conduct experiments on 3D posture tracking in occluded scenarios. Specifically, we introduce 4 distinct types of barriers between the mmWave radar and the users to simulate occlusions: a mural, a cloth screen, a whiteboard, and a projector screen. Figure 2.15 illustrates the scenarios involving four different barriers and the corresponding results obtained with $m^3Track$. For the mural and cloth screen, $m^3Track$ effectively tracks the 3D postures of all users, as seen in Fig. 2.15a and b. The average joint tracking errors for these two types of barriers are 45.7 and 44.2 mm, respectively,

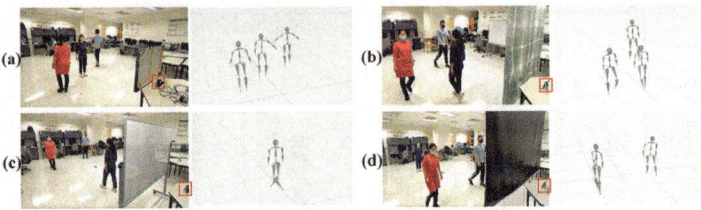

**Fig. 2.15** Multi-user 3D posture tracking in occluded scenarios

which are comparable to those observed in non-occluded environments. However, for the whiteboard and projector screen barriers, $m^3Track$ fails to accurately capture the details of closer users and loses the primary information of users further away, resulting in incomplete multi-user 3D posture tracking results, as shown in Fig. 2.15c and d. This issue arises because barriers with complex structures and materials that are difficult for signals to penetrate cause significant signal attenuation and phase shifts, preventing the precise measurement of user range, angle, and Doppler.

### 2.6.6 Performance of User Joining and Leaving

We evaluate the 3D posture tracking capabilities of $m^3Track$ in scenarios where users enter and exit the sensing area. In the experiment, two users are initially present, and a third user walks into the area before leaving again. Figure 2.16 presents the reconstructed 3D postures along with the corresponding video frames when a user joins and leaves the multi-user scenario. It shows that $m^3Track$ effectively reconstructs and tracks all users in the sensing area as new users enter and existing ones leave. Specifically, when a user enters the area, as shown in Fig. 2.16a and b, $m^3Track$ detects their presence at a given frame and begins tracking their posture. Later, as the user exits the area, as seen in Fig. 2.16c and d, $m^3Track$ loses track of the user at a particular frame. These results show that $m^3Track$ is capable

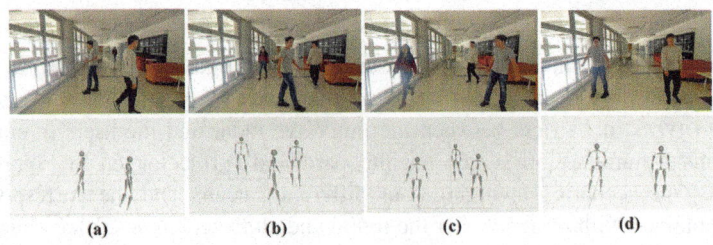

**Fig. 2.16** Multi-user 3D posture tracking for user joining and leaving

## 2.6 Evaluation

of accurately tracking multiple users in scenarios where users join and leave the sensing area.

### 2.6.7 Comparison with SOTA Systems

We evaluate the performance of the proposed system by comparing it with two state-of-the-art (SOTA) systems: *WiPose* [3] and *mm-Pose* [4]. *WiPose* utilizes commercially available WiFi devices to extract CSI data and 3D velocity profiles, which are then used to construct 3D skeletons for a single user through an RNN-based deep learning model. *mm-Pose* employs COTS mmWave radar to extract mmWave range-angle patterns, which are processed by a CNN-based model to estimate and generate a single user's skeleton. In addition to these two systems, there are other RF-based skeleton construction and tracking methods, such as *RF-Pose3D* [9], but due to the requirement of specialized hardware, a direct comparison with these systems is not feasible. Therefore, we focus on comparing with the two SOTA systems that operate on COTS devices. To conduct the comparison, we implement both *WiPose* and *mm-Pose* using a WiFi testbed and mmWave radar, respectively. Since both systems are designed for single-user scenarios, we compare the joint tracking errors using identical single-user experimental data. We present the average joint tracking errors of $m^3Track$ in multi-user scenarios for further comparison.

Figure 2.17 presents the joint tracking errors of *WiPose*, *mm-Pose*, and $m^3Track$ in single-user scenarios, as well as the performance of $m^3Track$ in multi-user scenarios. It shows that $m^3Track$ outperforms the other systems, achieving the lowest posture reconstruction errors. Specifically, $m^3Track$ achieves comparable results in both single-user and multi-user scenarios, with joint tracking errors significantly lower than those of *WiPose*, which has an average error of 38.7 mm.

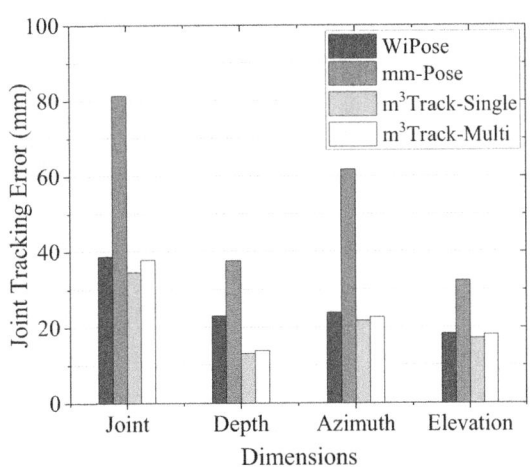

**Fig. 2.17** Joint tracking errors for $m^3Track$, *WiPose*, and *mm-Pose*

However, while *WiPose* is limited to single-user tracking, it extends its capability to handle multi-user scenarios effectively. For the COTS mmWave-based method *mm-Pose*, the joint tracking errors are relatively high, with an error of 81.3 mm for joints and other dimensions showing errors of 37.6, 61.8, and 32.4 mm. This can be attributed to the fact that *mm-Pose* uses a CNN to extract spatial features but does not account for temporal dependencies across different time points. In contrast, $m^3$*Track* integrates convolutional operations with temporal relations, allowing it to deliver more accurate multi-user 3D posture tracking while using the same COTS mmWave radar.

### 2.6.8 Localization Performance

Since $m^3$*Track* continuously localizes all users and tracks their walking trajectories in 3D space, it is essential to evaluate its localization performance for multiple users. To ensure a comprehensive evaluation, experiments are conducted in both an indoor environment (the lab) and an outdoor setting (depicted in Fig. 2.13f). During the experiment, we compute the geometric center of the reconstructed postures in 3D space as the predicted location for each user. As accurately obtaining user positions from video frames is challenging, the ground truth locations are marked on the floor, and the users follow these predefined trajectories during the evaluation. Figure 2.18 presents the cumulative distribution function (CDF) of localization errors for different user counts in two distinct environments. It is evident that the number of users and the environmental conditions have small influence on the localization accuracy of $m^3$*Track*. In particular, for single-user scenarios, $m^3$*Track* achieves a median localization error of 18.9 mm with a standard deviation of 11.4 mm, while for four-user scenarios, the median error is 21.5 mm with a standard

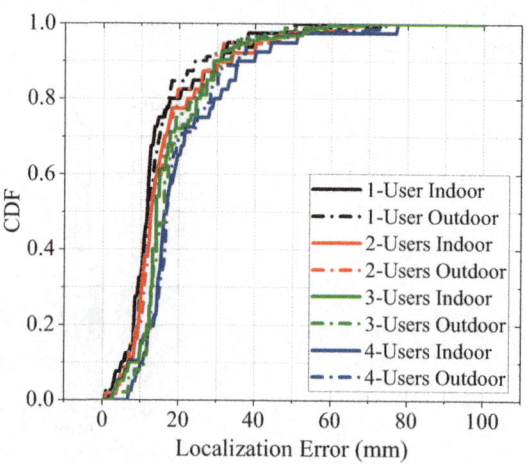

**Fig. 2.18** Localization errors in different environments

deviation of 13.5 mm, showing only slight variations. In comparison, *RF-Pose3D* [9], which utilizes specially-designed RF radar, reports localization errors of 17, 28, and 23 mm along the X, Y, and Z axes for multiple users. Our system, however, achieves comparable localization performance using only a COTS mmWave radar. These results demonstrate that $m^3$*Track* is capable of accurately localizing each user and continuously tracking their trajectory in multi-user environments.

### 2.6.9 Impact of Distance between Users

To further investigate the effect of the user distance on separating and tracking multiple users' postures, we analyze the performance with and without the use of the MVDR technique. The MVDR method is applied to enhance the signal-to-noise ratio, which aids in refining the angle estimation and thus enables better user separation. To evaluate this, we compare the results with and without MVDR under varying user distances. Figure 2.19 illustrates examples of 3D posture reconstruction for two users, both standing closely with their arms adjacent, under two conditions: with and without the MVDR technique. It shows that when the MVDR approach is applied, $m^3$*Track* effectively distinguishes the two users and accurately reconstructs their individual postures. In contrast, without the MVDR method, the users' postures are not well-separated, particularly in the adjacent body parts. These results demonstrate that the MVDR-based technique enables reliable separation and tracking of users even when they are positioned close to one another.

In addition, we evaluate the user detection recall, which quantifies the proportion of detected and separated users relative to the total number of users. This metric helps evaluate $m^3$*Track*'s performance in distinguishing all users without missing any, across different user distances. The distance between users is defined as the linear distance between their closest body parts. Figure 2.20a presents the user detection recall for varying distances and different numbers of users. As the distance between users narrows, the recall decreases, indicating that smaller distances complicate the separation of multiple users. Nevertheless, when the MVDR technique is applied, $m^3$*Track* can still accurately separate users, achieving

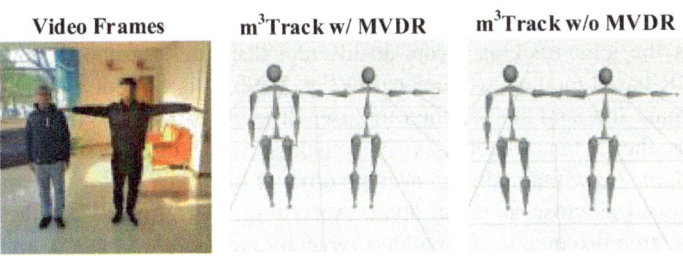

**Fig. 2.19** Posture tracking in adjacent distances

**Fig. 2.20** Performance under different user distances. (**a**) User detection recall. (**b**) Joint tracking errors

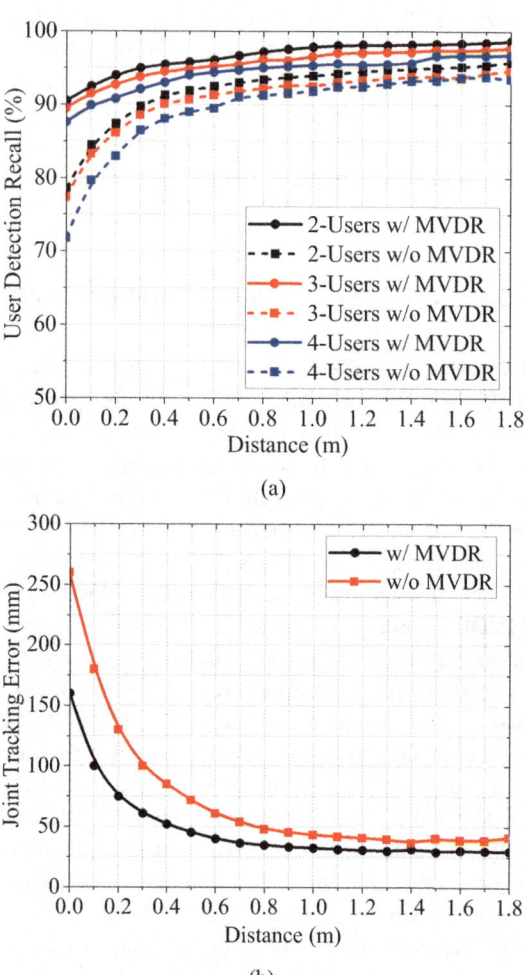

average recall rates of 90.5, 89.5, and 87.6% for 2-user, 3-user, and 4-user scenarios, respectively, even when users are positioned next to each other. This highlights the robustness of $m^3Track$ in detecting and separating multiple users. Additionally, we assess the joint tracking errors at different distances, both with and without the MVDR-based method, as depicted in Fig. 2.20b. It is evident that the tracking errors remain low and stable when the user distance exceeds 0.6 m, but increase sharply as the distance decreases. Nonetheless, when the MVDR technique is employed, $m^3Track$ maintains an average error of only 51.7 mm, even when users are positioned as close as 0.4 m apart. According to [21], the ideal face-to-face communication distance for individuals typically ranges between 0.46 and 1.22 m. Therefore, $m^3Track$ is capable of reliably tracking the postures of multiple users within most practical communication distances.

**Table 2.2** Joint tracking errors of different distances to radar

| Distance(m) | 1.2–2.0 | 2.0–3.0 | 3.0–4.0 | 4.0–5.0 | 5.0–6.0 | 6.0–7.0 |
|---|---|---|---|---|---|---|
| Error(mm) | 39.3 | 38.4 | 40.1 | 42.0 | 48.6 | 58.7 |

## 2.6.10 Impact of Distance to Radar

We perform a quantitative evaluation of the joint tracking errors in posture tracking at varying distances between users and the radar. For this experiment, we use data from 4-user scenarios within the experimental space ranging from 1.2 to 7 m, recording the straight-line distance from each user to the radar, and then measuring the joint tracking errors for each user across different distance ranges. Table 2.2 presents the average joint tracking errors for each user at these different distances from the radar. As observed, the joint tracking errors show a slight increase with increasing distance, primarily due to the attenuation of mmWave signals over longer distances. However, the difference between the minimum and maximum joint tracking errors across the distance ranges is only about 20 mm on average. This result highlights that $m^3Track$ exhibits minimal sensitivity to variations in the distance from the radar.

## 2.7 Summary

In this chapter, we propose $m^3Track$, a system that leverages a single COTS mmWave radar to enable 3D posture tracking in multi-user scenarios. It first detects and isolates all users present in the mmWave signals. Then, a novel deep learning model is designed to reconstruct the 3D posture of each individual user. Afterward, the reconstructed 3D postures are mapped into real-world 3D space for tracking the positions of all users. Therefore, $m^3Track$ achieves practical and accurate multi-user 3D posture reconstruction and tracking. Extensive experiments conducted in real-world multi-user scenarios validate the system's accuracy and robustness.

## References

1. Jiang, W., Yin, Z.: Human activity recognition using wearable sensors by deep convolutional neural networks. In: Proceedings of the ACM MM '15, pp. 1307–1310. Brisbane (2015)
2. Wang, J., Huang, Z., Zhang, W., Patil, A., Patil, K., Zhu, T., Shiroma, E.J., Schepps, M.A., Harris, T.B.: Wearable sensor based human posture recognition. In: Proceedings of the IEEE Big Data '16, pp. 3432–3438. IEEE, Washington DC (2016)
3. Jiang, W., Xue, H., Miao, C., Wang, S., Lin, S., Tian, C., Murali, S., Hu, H., Sun, Z., Su, L.: Towards 3d human pose construction using wifi. In: Proceedings of the ACM MobiCom '20, pp. 23:1–23:14. London (2020)

4. Sengupta, A., Jin, F., Zhang, R., Cao, S.: mm-pose: real-time human skeletal posture estimation using mmwave radars and cnns (2019). CoRR abs/1911.09592
5. Sengupta, A., Jin, F., Cao, S.: Nlp based skeletal pose estimation using mmwave radar point-cloud: A simulation approach. In: Proceedings of the IEEE RadarConf '20, pp. 1–6. Florence (2020)
6. Xue, H., Ju, Y., Miao, C., Wang, Y., Wang, S., Zhang, A., Su, L.: mmmesh: towards 3d real-time dynamic human mesh construction using millimeter-wave. In: Proceedings of the ACM MobiSys '21, pp. 269–282. Wisconsin, USA (2021)
7. Zhao, P., Lu, C.X., Wang, J., Chen, C., Wang, W., Trigoni, N., Markham, A.: mid: Tracking and identifying people with millimeter wave radar. In: Proceedings of the IEEE DCOSS '19, pp. 33–40. Santorini (2019)
8. Wu, C., Zhang, F., Wang, B., Liu, K.J.R.: mmtrack: passive multi-person localization using commodity millimeter wave radio. In: Proceedings of the IEEE INFOCOM '20, pp. 2400–2409. IEEE, Toronto (2020)
9. Zhao, M., Tian, Y., Zhao, H., Alsheikh, M.A., Li, T., Hristov, R., Kabelac, Z., Katabi, D., Torralba, A.: Rf-based 3d skeletons. In: Proceedings of the ACM SIGCOMM '18, pp. 267–281. Budapest (2018)
10. Zhao, M., Liu, Y., Raghu, A., Li, T., Zhao, H., Torralba, A., Katabi, D.: Through-wall human mesh recovery using radio signals. In: Proceedings of the IEEE/CVF ICCV '19, pp. 10113–10122. Seoul (2019)
11. Zhao, M., Li, T., Abu Alsheikh, M., Tian, Y., Zhao, H., Torralba, A., Katabi, D.: Through-wall human pose estimation using radio signals. In: Proceedings of the IEEE CVPR '18, pp. 7356–7365. Salt Lake City (2018)
12. Habets, E.A.P., Benesty, J., Cohen, I., Gannot, S., Dmochowski, J.: New insights into the mvdr beamformer in room acoustics. IEEE Trans. Audio Speech Language Process. **18**(1), 158–170 (2009)
13. Xingjian, S., Chen, Z., Wang, H., Yeung, D.Y., Wong, W.K., Woo, W.C.: Convolutional lstm network: a machine learning approach for precipitation nowcasting. In: Proceedings of the NeurIPS '15, pp. 802–810. Montreal (2015)
14. Girshick, R.: Fast r-cnn. In: Proceedings of the IEEE ICCV '15, pp. 1440–1448 (2015)
15. Rohling, H.: Radar cfar thresholding in clutter and multiple target situations. IEEE Trans. Aerospace Electron. Syst. (4), 608–621 (1983)
16. Likas, A., Vlassis, N., Verbeek, J.J.: The global k-means clustering algorithm. Pattern Recogn. **36**(2), 451–461 (2003)
17. Ribeiro, M.I.: Kalman and extended kalman filters: Concept, derivation and properties. Inst. Syst. Robot. **43**, 46 (2004)
18. Instruments, T.: Awr1443 single-chip 76-ghz to 81-ghz automotive radar sensor integrating mcu and hardware accelerator (2021) [Online]. Available: https://www.ti.com/product/AWR1443
19. Instruments, T.: Dca1000evm real-time data-capture adapter for radar sensing evaluation module (2021) [Online]. Available: https://www.ti.com/tool/DCA1000EVM
20. Microsoft: Kinect for windows (2021) [Online]. Available: https://developer.microsoft.com/en-us/windows/kinect/
21. Sorokowska, A., Sorokowski, P., Hilpert, P., Cantarero, K., Frackowiak, T., Ahmadi, K., Alghraibeh, A.M., Aryeetey, R., Bertoni, A., Bettache, K., et al.: Preferred interpersonal distances: a global comparison. J. Cross-Cultural Psychol. **48**(4), 577–592 (2017)

# Chapter 3
# mmWave-based Facial Expression Reconstruction

**Abstract** In recent times, there has been a rapid rise in the market for facial expression reconstruction technologies, which enable a wide range of face-driven applications, such as VR modeling, human-computer interaction, and affective computing. Existing methods for 3D facial reconstruction rely heavily on cameras, which can raise privacy concerns and struggle in scenarios with obstructions or poor lighting. In this chapter, we introduce a passive and privacy-conscious 3D facial expression reconstruction framework, *mm3DFace*, which utilizes a mmWave radar to reconstruct dynamic facial expressions. *mm3DFace* processes the pre-captured mmWave radar signals and extracts key facial geometric features that reflect subtle facial changes using a ConvNeXt model with triple loss embedding. Following this, *mm3DFace* derives robust facial shapes that are invariant to distance and orientation leveraging 68 facial landmarks through a region-divided affine transformation. Subsequently, *mm3DFace* enhances the reconstructed facial expressions by applying a region-based amplification technique and then generates 3D avatars that represent the facial expressions. Evaluations in real-world environments involving 15 participants demonstrate that *mm3DFace* can accurately track 68 facial landmarks with an average normalized mean error of 3.94%, a mean absolute error of 2.30 mm, and a 3D mean absolute error of 4.10 mm. These results validate the effectiveness and practicality of the system for real-world facial expression reconstruction applications.

**Keywords** Millimeter wave · Facial expression reconstruction · Affine transformation · 3D facial avatar

## 3.1 Introduction

As a key medium for human emotional communication and social interaction, the human face holds significant importance in the physical world. The process of facial expression reconstruction refers to the technology used to create dynamic representations of faces that capture facial expressions, which has garnered substantial interest in recent years. With the rise of IoT devices, the scope of facial expression reconstruction has expanded significantly, covering diverse areas such as human-computer interaction, healthcare monitoring, VR modeling, and more. Additionally, facial reconstruction techniques play a crucial role in the field of affective computing [1], enabling systems to accurately perceive and appropriately react to an individual's emotional state across various application domains.

In order to implement affective computing across a wide array of contexts, facial expression reconstruction must be both pervasive and unobtrusive. For instance, considering the growing concerns over personal privacy and the common occurrence of mask-wearing, there is a significant demand for privacy-preserving, yet effective, facial expression reconstruction techniques. An example of such an application is analyzing users' preferences for advertisements based on their facial expressions in front of billboards. Furthermore, a facial expression reconstruction method that is robust to varying lighting conditions and nonintrusive could significantly contribute to safety-oriented affective computing, such as monitoring driver fatigue in challenging driving conditions. However, current camera-based facial expression reconstruction methods [2–4] often compromise privacy by revealing sensitive information, such as the surrounding environment and other people, and are prone to performance degradation in poor lighting or NLOS scenarios. Some recent approaches employ wearable devices for facial reconstruction [5–7], but these systems require users to wear specialized devices capable of sensing magnetic fields, capacitance, or electromyography, leading to intrusive user experiences and high costs. Additionally, such systems cannot passively reconstruct facial features without the user's active involvement. As a result, these existing solutions fall short of achieving truly ubiquitous affective computing.

In recent years, mmWave signals have found diverse applications beyond communications, extending to various sensing tasks such as localization [8, 9], activity recognition [10, 11], pose reconstruction [12–14], and more. Sensing using mmWave signals offers robustness under various lighting conditions and ensures privacy preservation. It has the added advantage of penetrating barriers such as masks, broadening its potential compared to camera-based methods. Furthermore, as a contactless sensing technique, mmWave provides a less intrusive user experience when compared to wearable devices. Consequently, the versatility of mmWave signals motivates their use in developing a nonintrusive facial reconstruction system, which can be easily integrated into real-world scenarios for affective computing.

However, achieving nonintrusive facial expression reconstruction using mmWave signals presents several practical challenges. Firstly, because facial expressions are tiny changes, it is crucial to effectively extract features from the mmWave data that capture these tiny changes. Secondly, human faces may position at varying distances and orientations relative to the mmWave radar in real-world settings, which requires accommodating such variations. Lastly, to facilitate practical applications, we must develop the capability to generate 3D facial models that can continuously reflect dynamic facial expressions derived from facial landmarks.

To achieve nonintrusive facial expression reconstruction, we present *mm3DFace*, which employs a COTS mmWave radar to generate 3D human facial models that

## 3.1 Introduction

dynamically reflect facial expressions. Initially, *mm3DFace* uses mmWave signals to detect a user's face and preprocesses the signals to extract facial features. Using the pre-processed mmWave data, *mm3DFace* employs a ConvNeXt model to derive feature representations, followed by the extraction of facial geometric characteristics that capture subtle expression variations via a specially designed triple loss embedding. Subsequently, *mm3DFace* reconstructs the facial shape by identifying 68 facial landmarks and applying affine transformations to different regions of the face, ensuring robustness to variations in distance and orientation. With these reconstructed facial shapes, *mm3DFace* further refines the facial expressions through a custom regional amplification technique. Finally, 3D facial avatars that continuously adapt to facial expressions are generated using the FLAME model. We evaluate *mm3DFace* through a series of comprehensive experiments in real-world settings. The experiments demonstrate that *mm3DFace* effectively reconstructs 3D human faces with varying expressions. A visual representation of the *mm3DFace* system is provided in Fig. 3.1.

We summarize the key contributions of our work as follows:

- We introduce a nonintrusive 3D facial expression reconstruction framework, *mm3DFace*, which utilizes mmWave signals to reconstruct 3D human faces that dynamically reflect facial expressions. To the best of our knowledge, *mm3DFace* is the first solution for 3D facial reconstruction based on radio frequency signals.
- We develop a ConvNeXt-based model integrated with a triple loss embedding to effectively capture facial geometric features that are sensitive to subtle changes in facial expressions.
- We propose affine transformation and regional amplification to reconstruct facial expressions that are robust to variations in distance and orientation, using 68 facial landmarks, and to generate 3D avatars that continuously reflect facial expressions.

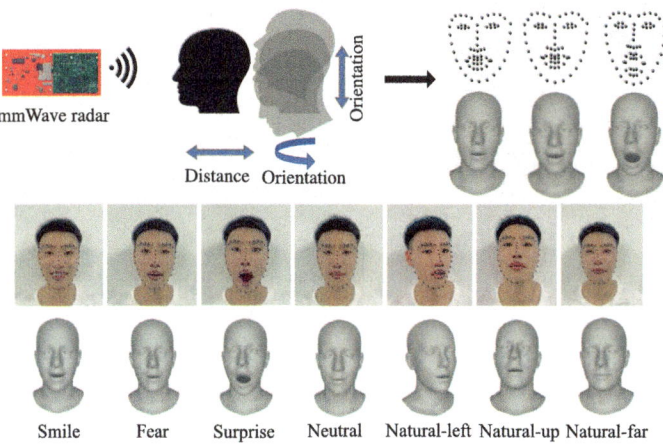

**Fig. 3.1** Illustration of *mm3DFace* system

- We perform experiments involving 15 participants in real-world settings. The results demonstrate that *mm3DFace* effectively reconstructs facial expressions, achieving a normalized mean error of 3.94%, a mean absolute error of 2.30 mm for tracking human faces with 68 landmarks, and a mean absolute error of 4.10 mm for 3D facial reconstruction.

## 3.2 System Overview

To enable nonintrusive 3D facial reconstruction in practical environments, we introduce *mm3DFace*, a system that utilizes mmWave signals to continuously generate 3D facial expressions. The architecture of *mm3DFace* is illustrated in Fig. 3.2.

To reconstruct 3D facial expressions, *mm3DFace* first acquires mmWave signals emitted by a radar and reflected from the human face, followed by preprocessing these signals to extract relevant facial features. Next, *mm3DFace* utilizes a ConvNeXt model to derive facial feature representations, and further refines these features to capture fine-grained changes in facial expressions through a specially designed triple loss embedding. To perform 3D facial reconstruction using the extracted facial geometric features, *mm3DFace* applies affine transformations across different facial regions to reconstruct facial shapes with 68 landmarks, ensuring robustness against variations in distance and orientation. Subsequently, *mm3DFace* reconstructs facial expressions using the 68 landmarks through a custom regional amplification technique, which emphasizes subtle facial changes to preserve detailed expressions. Leveraging both the facial shapes and expressions, *mm3DFace* generates continuous 3D facial avatars using the FLAME model.

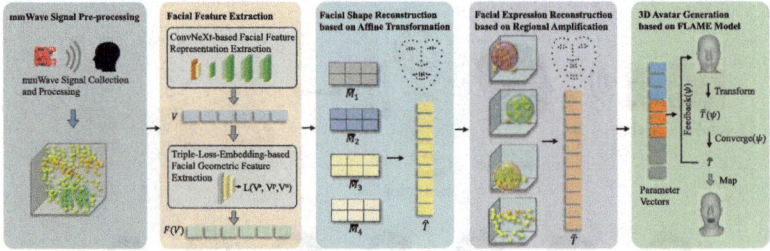

**Fig. 3.2** System framework of *mm3DFace*, which can reconstruct 3D facial expressions leveraging mmWave signals

## 3.3 mmWave Signal Pre-processing

*mm3DFace* first processes mmWave signals to extract facial characteristics before performing 3D facial expression reconstruction.

Using a COTS mmWave radar, *mm3DFace* employs FMCW techniques to generate chirps with a linearly increasing frequency from the radar's transmitting antennas. These signals are reflected off a human face and then captured by the receiving antennas of the radar. To extract the sensing data of the human face, *mm3DFace* combines the transmitted and received signals to compute the IF signals. After acquiring the reflected IF signals from the mmWave, *mm3DFace* proceeds to compute both the ranges and angles in order to extract facial features for 3D facial expression reconstruction. The distance to an object, denoted as $r$, can be expressed by the formula $r = \frac{cfT_c}{2B}$, indicating a direct proportionality between the object's range $r$ and the frequency $f$ of the IF signal. By applying a range-FFT to the IF signal, *mm3DFace* is capable of determining the range and generating *Range-Spectrum*. *mm3DFace* calculates the AoA of the human face using a combination of multiple transmit and receive antennas. The angles of the object relative to the mmWave radar—comprising the azimuth angle $\theta$ and the elevation angle $\phi$—are given by the equations $\theta = \sin^{-1}\left(\frac{\lambda \omega_a}{2\pi d_1}\right)$ and $\phi = \sin^{-1}\left(\frac{\lambda \omega_e}{2\pi d_2}\right)$, where $d_1$ and $d_2$ are the physical separations of the receive antennas in the azimuth and elevation directions, respectively, and $\omega_a$ and $\omega_e$ are the phase shifts in the MIMO channels along the azimuth and elevation. Additionally, *mm3DFace* leverages a TDM-MIMO scheme to compute the angles with virtual antenna arrays. After the AoA computation, *Azimuth-Spectrum* and *Elevation-Spectrum* are produced, which provide information about the azimuth and elevation angles, respectively.

The *Range-Spectrum*, *Azimuth-Spectrum*, and *Elevation-Spectrum* represent human faces in the three-dimensional space (i.e., X, Y, and Z axes). As illustrated in Fig. 3.3, a human face is detected by the mmWave radar. From the range (X) and elevation (Z) viewpoint, *mm3DFace* captures the depth information of the face in the *Range-Elevation-Spectrum*, which manifests as a circular region of high intensity corresponding to the facial profile from a side view. From the azimuth (Y) and elevation (Z) viewpoint, *mm3DFace* gathers the planar characteristics of the face in the *Azimuth-Elevation-Spectrum*, represented by a dispersed region of high intensity corresponding to the frontal view of the face. By integrating range, elevation, and azimuth, *mm3DFace* constructs the *Range-Angle-Spectrum*, which are three-dimensional matrices that encode the full 3D features of the human face. To optimize the utility of the pre-processed data, the *Range-Angle-Spectrum* are filtered to remove weak components and are then reshaped into 3D data cubes for subsequent analysis.

Building on the pre-processed mmWave signals, *mm3DFace* subsequently extracts facial characteristics for mmWave-based 3D facial expression reconstruction.

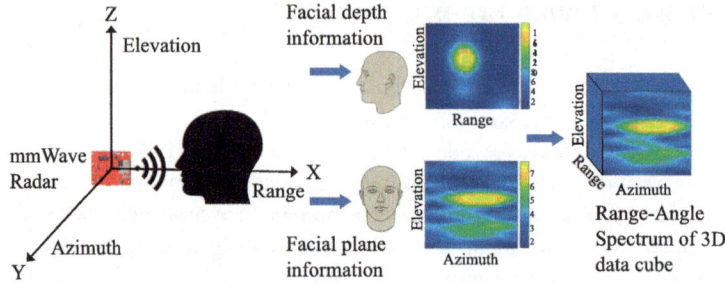

**Fig. 3.3** Illustration of signal pre-processing

## 3.4 Facial Feature Extraction

Following the pre-processing of the mmWave signals, *mm3DFace* proceeds to extract facial geometric characteristics for 3D facial reconstruction using a ConvNeXt model with a triple loss embedding, as depicted in Fig. 3.4.

Through the processing of mmWave signals, *mm3DFace* generates *Range-Angle-Spectrum*. These *Range-Angle-Spectrum* represent facial characteristics in the 3D space, i.e., they illustrate the spatial relationships of a user's face. Therefore, it is appropriate to apply a CNN to extract these spatial relationships as facial feature representations. *mm3DFace* utilizes a ConvNeXt model [15] to derive facial feature representations from the *Range-Angle-Spectrum*. A *Range-Angle-Spectrum* is represented by $(\theta, \phi, r)$, where $\theta$ is the azimuth angle, $\phi$ is the elevation angle, and $r$ is the range. This spectrum is fed into the ConvNeXt model, where convolutional operations are performed in the plane defined by $\theta$ and $\phi$.

The ConvNeXt model designed for this task comprises four hidden layers dedicated to feature extraction. Through these four layers, convolutional operations are applied to the plane formed by $\theta$ and $\phi$ in order to reduce the spatial dimensions while preserving the essential facial information. Simultaneously, the dimensionality of the range is progressively expanded to filter out irrelevant noise from other ranges, focusing solely on the face. The computational ratio across the four hidden layers is empirically set to 3:3:15:3 in the proposed ConvNeXt model. Following the ConvNeXt processing, facial feature representations are extracted

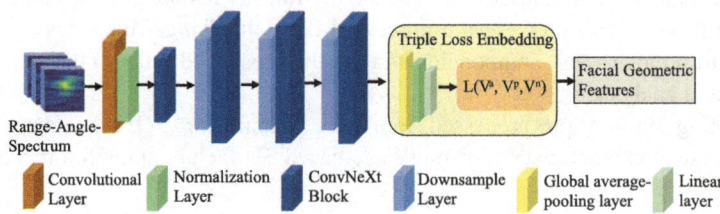

**Fig. 3.4** Architecture of ConvNeXt model with triple loss embedding

## 3.4 Facial Feature Extraction

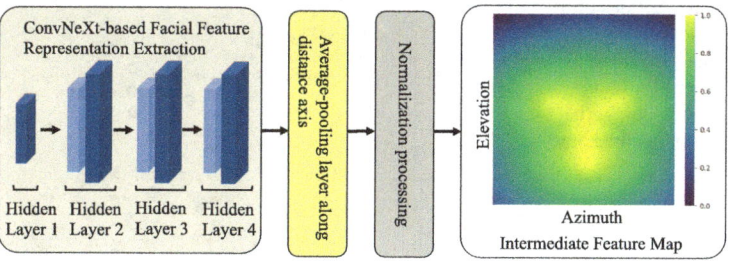

**Fig. 3.5** The intermediate feature map of ConvNeXt

from the *Range-Angle-Spectrum*. Figure 3.5 illustrates an intermediate feature map produced by the ConvNeXt model, where the facial feature representations of the human face are passed through an average-pooling layer along the range axis and a normalization layer for visualization. We can see from the feature map that it highlights the approximate spatial location of the face and the spatial relationships between facial components. Therefore, the facial feature representations extracted by the ConvNeXt model progressively refine their alignment with human faces, derived from the pre-processed mmWave signals, which are subsequently used for 3D facial expression reconstruction.

To explore the feasibility of differentiating between various facial expressions using facial feature representations, we conduct an initial experiment involving 5 volunteers, asking them to repeatedly perform 3 distinct expressions (i.e., neutral, smile, and surprise) in front of a mmWave radar. Once the facial feature representations for each expression are extracted, principal component analysis (PCA) is applied to the facial features to visualize the similarity distribution between the expressions. Figure 3.6a illustrates the distribution of facial feature representations for the 3 expressions across the 5 volunteers, where the first two principal components of the facial features are plotted in a 2D space. The preliminary results indicate that identical expressions tend to form distinct clusters, while different expressions are separated into separate clusters, suggesting that mmWave signals can effectively capture facial movements. However, the distinction between different expressions is not clearly marked, as some distinct expressions are positioned closer to each other than similar ones in the distribution. This shows that only using facial feature representations cannot show subtle differences in facial expressions. The primary reason is the resolution constraints and the error-prone characteristics of COTS mmWave radars. Given the mmWave signals emitted by COTS devices, different facial expressions may produce similar feature representations, making them challenging to differentiate. Therefore, facial feature representations are insufficient for accurately recognizing and capturing the subtle variations in facial expressions.

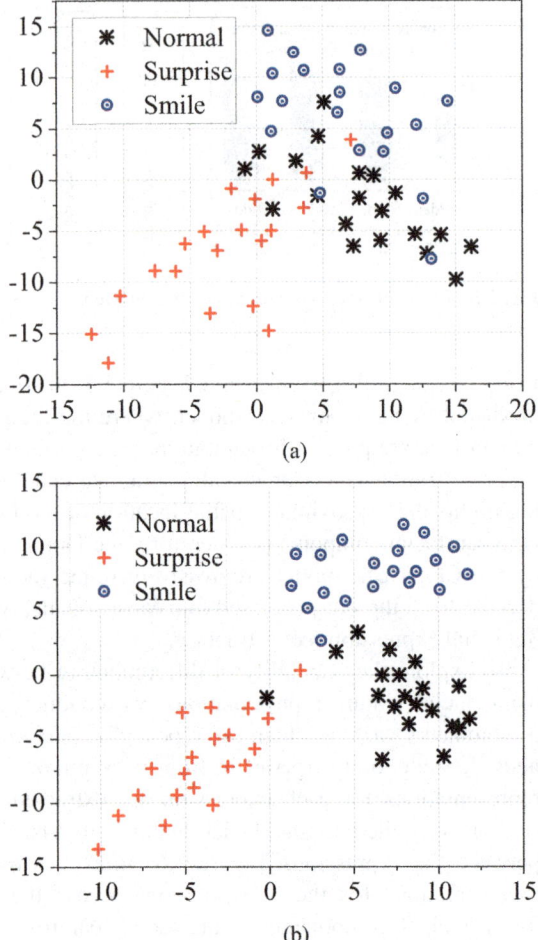

**Fig. 3.6** Distributions of 3 facial expressions with/without triple loss embedding. (**a**) Without triplet loss embedding. (**b**) With triplet loss embedding

### 3.4.1 Triple-Loss-Embedding-based Facial Geometric Feature Extraction

The resolution constraints of mmWave signals impede the direct detection of subtle variations in facial expressions. To address this limitation in mmWave signal resolution when sensing human faces, we propose increasing the separation between different facial expressions at the feature level, which facilitates the distinction of various facial expressions despite the resolution limitations of the mmWave signals.

A triple-loss embedding approach [16] is introduced to extract facial geometric features, aimed at enhancing the distinction between different facial expressions at the feature level. This method maps facial feature representations into facial geometric features, capturing fine-grained facial details and subtle variations. The

## 3.4 Facial Feature Extraction

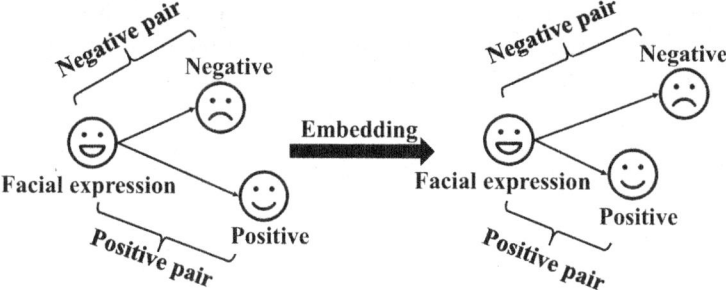

**Fig. 3.7** Illustration of triple loss embedding

core concept involves projecting identical facial expressions into a single embedding space, while distinct facial expressions are mapped to separate embedding spaces. As illustrated in Fig. 3.7, the distance between two facial expressions represents their degree of similarity. A larger distance indicates greater differences (i.e., negative pairs), while a smaller distance indicates higher similarity (i.e., positive pairs) between the expressions. Through the triple-loss embedding, *mm3DFace* reduces the distance between similar facial expressions and increases the distance between different ones. Reducing the distance for similar expressions helps extract common features without being dependent on specific users, thus ensuring better generalization across different faces. Increasing the distance between distinct facial expressions allows for a clearer differentiation of facial features under various expressions, even with resolution-limited mmWave signals.

In the triple-loss embedding, *mm3DFace* initially employs a global average-pooling layer to convert the three-dimensional facial feature representations into a one-dimensional feature vector $V$, effectively reducing the dimensionality to match the input requirements of the triple-loss embedding. The output of the triple-loss embedding is denoted as $F(V) \in \mathbb{R}^d$, with the constraint that $\|F(V)\|_2 = 1$, where the feature vector $V$ is mapped to a $d$-dimensional Euclidean space. The objective of the triple-loss embedding is to ensure that the feature vector $V_i^a$ corresponding to a specific facial expression is closer to all positive feature vectors $V_i^p$ (i.e., $V_i^p$ is identical to $V_i^a$) than to all negative feature vectors $V_i^n$ (i.e., $V_i^n$ differs from $V_i^a$), such that:

$$\|V_i^a - V_i^p\|_2^2 + \gamma < \|V_i^a - V_i^n\|_2^2, \quad \forall (V_i^a, V_i^p, V_i^n) \in \mathcal{T}, \tag{3.1}$$

where $\gamma$ represents the margin that enforces a minimum distance between positive and negative pairs, and $\mathcal{T}$ denotes the set of all possible triplets in the training dataset. The loss function is defined to minimize $L$, given by:

$$\arg\min_F L = \sum_i^N \left[ \max\left( \left\| F(V_i^a) - F(V_i^p) \right\|_2^2 - \left\| F(V_i^a) - F(V_i^n) \right\|_2^2 + \gamma, 0 \right) \right], \quad (3.2)$$

where $\max(\cdot)$ is used to prevent negative distance loss between pairs.

The triplets $(V^a, V^p, V^n)$ are derived from the training dataset, where numerous triplets can satisfy the constraint in Eq. (3.1), resulting in a loss function $L$ equal to zero. These triplets not only increase computational complexity but also hinder the convergence speed during model training. Therefore, only hard triplets that fail to meet the condition in Eq. (3.1) are selected for training. Since identical facial expressions from different individuals tend to have larger distances than those from the same individual, we define pairs of identical facial expressions from different users as hard positive pairs. *mm3DFace* prioritizes selecting a higher proportion of hard positive pairs $(V^a, V^p)$ and randomly chooses different facial expressions as negative pairs $(V^a, V^n)$ to form hard triplets. Through the design of the triple-loss embedding, the distances between facial features of both identical and distinct facial expressions are adjusted. This process generates facial geometric features capable of capturing subtle changes in facial expressions, even with resolution-limited mmWave signals. Figure 3.6b displays the distributions of facial geometric features for the three facial expressions across 5 volunteers after applying the triple-loss embedding. It is evident that the boundaries between different facial expressions are more pronounced, and identical expressions are clustered together. Thus, facial expressions captured by mmWave signals are more distinguishable at the feature level, which aids in overcoming the resolution limitations of the mmWave signals.

By utilizing the ConvNeXt model with triple-loss embedding, *mm3DFace* derives facial geometric features from mmWave signals for the purpose of 3D facial reconstruction.

## 3.5 3D Facial Reconstruction

Leveraging the extracted geometric characteristics of the face, *mm3DFace* then reconstructs both the facial geometry and the facial expressions, ultimately producing 3D facial avatars. *mm3DFace* utilizes 68 key facial landmarks [17] to model a human face.

## 3.5 3D Facial Reconstruction

### 3.5.1 Facial Shape Reconstruction

The 68 facial landmarks capture both the overall facial structure and the spatial relationship of facial contours and features. To reconstruct the facial shape using these landmarks, one simple approach is to predict the locations of the 68 points based on the facial geometric features $F(V)$. However, the relative distance and orientation of a human face to a radar are typically variable, which can influence the accuracy of the landmark prediction. For instance, a change in the face's distance from the radar results in scaling of the facial landmarks, while variations in the face's orientation cause rotation of the landmarks. Regressing directly from these scaled or rotated facial landmarks leads to incorrect placements of facial features and distortion of facial contours. Thus, it becomes crucial to account for both distance and orientation in order to achieve precise and reliable reconstruction of facial shapes using the 68 landmarks.

Affine transformations [18], as a method of geometric manipulation, are capable of zooming, rotating, translating, and skewing points within a plane, all while preserving their relative geometric relationships. To ensure accurate and reliable reconstruction of facial shapes, *mm3DFace* employs affine transformations to adjust the scaled or rotated facial landmarks resulting from variations in distance and orientation, thereby aligning them with the actual scene. Specifically, by applying affine transformations, the $i_{th}$ facial landmark $(x_i, y_i)$ is mapped to $(\tilde{x}_i, \tilde{y}_i)$, as expressed by the equation:

$$\begin{bmatrix} \tilde{x}_i \\ \tilde{y}_i \\ 1 \end{bmatrix} = \mathbf{M} \cdot \begin{bmatrix} x_i \\ y_i \\ 1 \end{bmatrix} = \begin{bmatrix} m_{11} & m_{12} & m_{13} \\ m_{21} & m_{22} & m_{23} \\ 0 & 0 & 1 \end{bmatrix} \cdot \begin{bmatrix} x_i \\ y_i \\ 1 \end{bmatrix}, \quad (3.3)$$

where $\mathbf{M}$ denotes the affine transformation matrix. In this matrix, the zoom and translation parameters are adjusted to compensate for the effects of distance, while the rotation and skew parameters are modified to account for the changes in orientation.

Typically, an affine matrix can be constructed by choosing three non-collinear points. However, the resulting affine matrices can vary considerably depending on the selection of the landmarks. For instance, the affine matrix generated by choosing three landmarks from the eyes differs substantially from that constructed using three landmarks from the mouth or nose.

To minimize the discrepancies between affine matrices, *mm3DFace* partitions the face into four distinct regions, as illustrated in Fig. 3.8, with each region having its own affine matrix. Within each region, any set of three non-collinear landmarks is used to generate a stable affine matrix, ensuring consistency across the matrices. The four regions, as shown in the figure, consist of the left eye and eyebrow, right eye and eyebrow, nose and contour, and the mouth. *mm3DFace* uses facial geometric

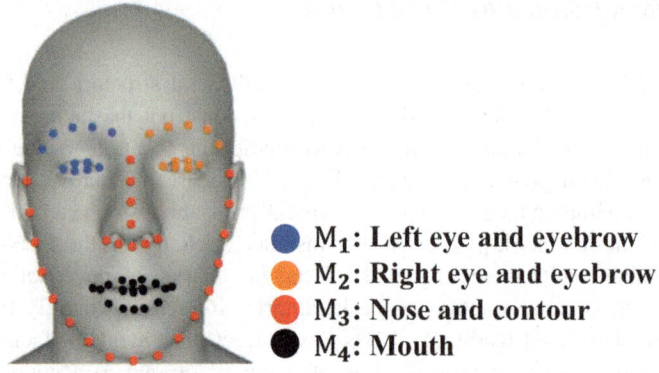

**Fig. 3.8** Region division for affine transformation

features $F(V)$ to derive four separate affine matrices through a linear mapping process. This can be expressed as

$$\hat{\mathbf{M}}_1, \hat{\mathbf{M}}_2, \hat{\mathbf{M}}_3, \hat{\mathbf{M}}_4 = G(FC(F(V), \Theta')), \tag{3.4}$$

where $\hat{\mathbf{M}}_1$, $\hat{\mathbf{M}}_2$, $\hat{\mathbf{M}}_3$, and $\hat{\mathbf{M}}_4$ are the affine matrices for the four regions, $FC(\cdot)$ represents a fully-connected layer that outputs a 24-dimensional vector, $\Theta'$ refers to the trainable parameters, and $G$ is the operation that maps the vector parameters into the four matrices. Once the four affine matrices are computed, the 68 landmarks' facial shapes, denoted as $\hat{T}$, are obtained by the following expression:

$$\hat{T} = Q(R_1(\hat{\mathbf{M}}_1), R_2(\hat{\mathbf{M}}_2), R_3(\hat{\mathbf{M}}_3), R_4(\hat{\mathbf{M}}_4)), \tag{3.5}$$

where $R_1(\cdot)$ is the affine transformation for the left eye and eyebrow landmarks, $R_2(\cdot)$ is the transformation for the right eye and eyebrow landmarks, $R_3(\cdot)$ is the transformation for the nose and contour landmarks, $R_4(\cdot)$ is the transformation for the mouth landmarks, and $Q(\cdot)$ is the aggregation function that combines the transformed landmarks from the four regions.

To compute the four affine matrices for the distinct regions, we define a loss function based on the discrepancy between the predicted affine matrices and the ground truth values (i.e., landmarks derived from images [19]). The loss function is formulated using the smooth $L_1$ loss [20], as follows:

$$L_P(\hat{\mathbf{M}}_i) = \begin{cases} \frac{(\mathbf{M}_i - \hat{\mathbf{M}}_i)^2}{2\delta} & \text{if } |\mathbf{M}_i - \hat{\mathbf{M}}_i| \leq \delta, \\ |\mathbf{M}_i - \hat{\mathbf{M}}_i| - \frac{1}{2}\delta & \text{if } |\mathbf{M}_i - \hat{\mathbf{M}}_i| > \delta. \end{cases} \tag{3.6}$$

where $\mathbf{M}_i$ represents the ground truth affine matrix for the $i_{th}$ region, and $\hat{\mathbf{M}}_i$ is the predicted affine matrix for the same region. To prevent issues with model

## 3.5 3D Facial Reconstruction

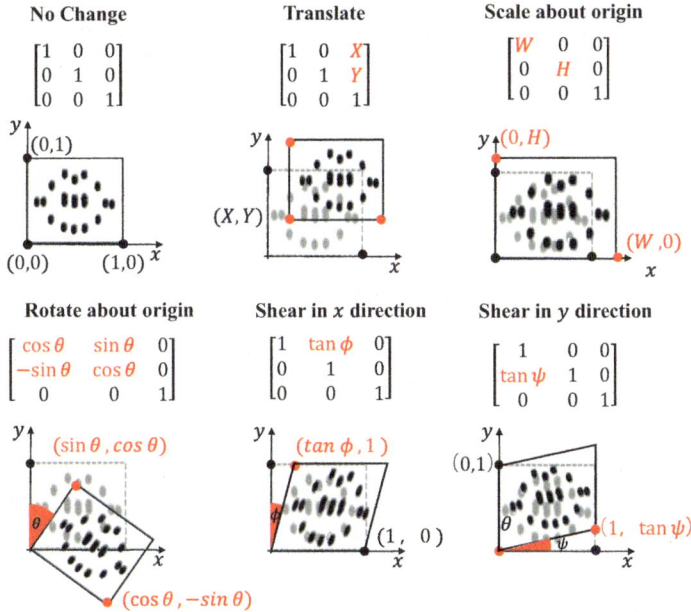

**Fig. 3.9** Illustration of affine transformation

convergence due to large penalty values for outliers, we introduce an outlier threshold, $\delta$, into the loss function to detect outlier parameters. As a result, the total loss for the four affine matrices can be expressed as:

$$L_o = \sum_{i=1}^{4} L_P(\hat{\mathbf{M}}_i). \tag{3.7}$$

Consider the affine matrix of the mouth region, denoted as $\mathbf{M}_4$. In Fig. 3.9, the affine transformation for the mouth region is illustrated with varying parameters for translation, scaling, rotation, and shearing of the landmarks. From the figure, it is evident that the values $m_{13}$ and $m_{23}$ (where $m_{ij}$ refers to the element in the $i_{th}$ row and $j_{th}$ column of the matrix) are responsible for translating the landmarks along the $X$ and $Y$ axes. Meanwhile, the elements $m_{11}$, $m_{12}$, $m_{21}$, and $m_{22}$ are adjusted to apply rotation to the landmarks within the $XY$ plane. As the affine matrices change, the landmarks of the mouth shift accordingly, allowing for accurate representation of various mouth shapes. Figure 3.10 presents the facial shapes generated by combining the landmarks from the four regions. It can be seen that the landmarks across different face orientations align closely with the ground truth data produced by Dlib [19], maintaining high accuracy and consistency.

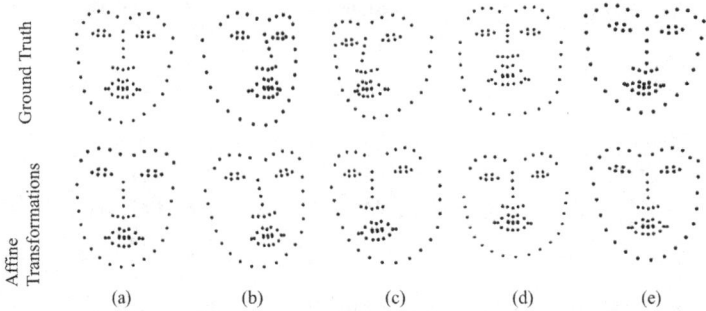

**Fig. 3.10** Examples of a user's facial shapes with 68 facial landmarks through affine transformation. (**a**) No toward. (**b**) Toward right. (**c**) Toward left. (**d**) Toward up. (**e**) Toward down

Leveraging the four affine matrices corresponding to distinct facial regions, *mm3DFace* effectively identifies facial landmarks and reconstructs facial shapes under varying face distances and orientations.

### 3.5.2 Facial Expression Reconstruction

Building upon the facial shapes defined by 68 landmarks, *mm3DFace* further reconstructs facial expressions. Facial expressions emerge from subtle variations in the appearance of different facial regions, so focusing on individual regions enhances the ability to capture the finer details of facial expressions. With this insight, instead of directly utilizing the facial shapes with 68 landmarks, *mm3DFace* exploits the sensing capabilities of mmWave signals to capture the varying appearances of different facial components, allowing for a more precise reconstruction of facial expressions. To achieve this, a regional amplification technique is developed to enhance the reconstruction of facial expression details.

The regional amplification approach leverages the unique properties of mmWave signals to partition the facial surface spatially and extract relevant geometric features for each section. By extracting these regional features, the method offers a finer resolution for individual facial areas, enhancing subtle variations in facial expressions even when working with low-resolution mmWave signals. The approach identifies the spatial regions corresponding to the left eye, right eye, and mouth within the mmWave signal and uses set subsections to locate additional facial areas. These facial region boundaries can be modeled as circles, with their centers and radii defined within the *Azimuth-Elevation-Spectrums* of mmWave signals. Using the derived affine transformation matrices, the center coordinates of these spatial regions are given by

$$(y_i, z_i) = \text{Trans}(\hat{\mathbf{M}}_i), \ i = 1, 2, 4, \tag{3.8}$$

## 3.5 3D Facial Reconstruction

where $\hat{\mathbf{M}}_i$ represents the affine matrix for the $i_{th}$ region, $(y_i, z_i)$ denotes the center coordinates in the *Azimuth-Elevation-Spectrums* of the mmWave signals, Trans($\cdot$) is the transformation function mapping affine matrix parameters to the center coordinates, and $i = 1, 2, 4$ corresponds to the regions of the left eye, right eye, and mouth, respectively.

As various facial areas reflect mmWave signals with distinct characteristics, the amplitude of the reflected signals varies across these regions. Hence, rather than using image processing techniques to detect regional boundaries, *mm3DFace* utilizes amplitude variations within each spatial region to estimate the radius of every facial area at the signal level. The radius for each region can be determined using the following equation:

$$\hat{r}_i = \max r_{i,j}$$

$$\text{subject to } \|E_m(y_i, z_i, r_{i,j}) - E_m(y_i, z_i, r_{i,j-1})\| \leq \varepsilon, \tag{3.9}$$

$$r_{i,j} = r_{i,j-1} + \Delta r, \ i = 1, 2, 4; \ j = 1, 2, 3, \ldots,$$

where $E_m(y_i, z_i, r_{i,j})$ represents the average amplitude of the mmWave signals at the center coordinates $(y_i, z_i)$ and a given radius $r_{i,j}$, $j$ denotes the index for incrementing the radius, $\Delta r$ is the radial step size, $\varepsilon$ is the threshold for amplitude variation, and $\hat{r}_i$ indicates the radius of the $i_{th}$ region.

Given the spatial bounds (i.e., center and radius), the mmWave signals corresponding to each facial region can be represented as

$$v_i = \mathrm{U}(y_i, z_i, \hat{r}_i), \ i = 1, 2, 4, \ v = v_1 \cup v_2 \cup v_3 \cup v_4, \tag{3.10}$$

where U($\cdot$) denotes the operator that extracts the mmWave signals within the spatial bounds of a given region, $v_i$ signifies the signal amplitude for the $i_{th}$ region, $v$ represents the overall signal amplitude across the entire face, and $v_3$ can be determined by the set operation, i.e., $v_3 = v - v_1 - v_2 - v_4$. Once the spatial scope of mmWave signals for each region has been established, *mm3DFace* proceeds to isolate regional features by focusing on the specific area of interest and excluding the contributions from other regions. This strategy is designed to emphasize even the slight variations in facial appearance. *mm3DFace* achieves this by masking the non-target regions with zero values and feeding the mmWave signals of the target region into the neural network. Consequently, the geometric features of each facial region are represented as

$$S_i = F(G'(v_i)), \ i = 1, 2, 3, 4, \tag{3.11}$$

where $G'(v_i)$ denotes the masking operation that preserves the mmWave signals of the $i_{th}$ region and sets the other regions to zero, $F(\cdot)$ represents the triple-loss embedding function, and $S_i$ denotes the geometric features extracted from the $i_{th}$ region.

Using the regional facial geometric features $S_i$, the appearance of each facial region can be represented as

$$H_i = FC_i(S_i), \quad i = 1, 2, 3, 4, \tag{3.12}$$

where $H_i$ denotes the output corresponding to the appearance of the $i_{th}$ facial region, along with its associated landmarks, and $FC_i(\cdot)$ is the fully connected layer applied to the $i_{th}$ region. By combining the appearance features of all facial regions, the set of 68 facial landmarks, which capture facial expressions, can be represented as

$$T' = \bar{G}(H_1, H_2, H_3, H_4), \tag{3.13}$$

where $T'$ is the final output representing facial expressions through 68 landmarks, and $\bar{G}(\cdot)$ is the operation that consolidates the landmarks from the four facial regions.

Based on the facial structures, the final facial landmarks are expressed as

$$\tilde{T} = \hat{T} + T', \tag{3.14}$$

where $\hat{T}$ refers to the facial structure with 68 landmarks, and $T'$ represents the facial expression with 68 landmarks. The loss function for generating facial expressions can be defined as

$$L_d = \begin{cases} \frac{(T-\tilde{T})^2}{2\delta} & \text{if } |T - \tilde{T}| \leq \delta, \\ |T - \tilde{T}| - \frac{1}{2}\delta & \text{if } |T - \tilde{T}| > \delta, \end{cases} \tag{3.15}$$

where $T$ denotes the ground truth of the 68 predicted facial landmarks, and $\tilde{T}$ represents the output of the 68 predicted facial landmarks. The overall loss function $L_g$ is given by

$$L_g = L_o + \alpha(t) L_d, \tag{3.16}$$

where $\alpha(t)$ is the fraction of training progress at time $t$, which adjusts the balance between the loss of facial structures and facial expressions as training advances.

By reconstructing facial expressions using the proposed regional amplification approach, *mm3DFace* derives the 68 facial landmarks that capture both the facial shapes and expressions.

### 3.5.3  3D Avatar Generation

To reconstruct 3D human faces from the 68 facial landmarks representing both shapes and expressions, *mm3DFace* utilizes FLAME (Faces Learned with an Artic-

## 3.5 3D Facial Reconstruction

ulated Model and Expressions) [21] to produce 3D facial avatars that dynamically convey facial expressions.

A FLAME model represents the human head with 5023 vertices, 4 joints (including the neck, jaw, and eyeballs), and blendshapes to explicitly generate a 3D model of head poses and facial expressions. It is primarily composed of four components: a template mesh to normalize face shapes, a shape blendshape to model identity-related shape variations, a pose blendshape to adjust for pose deformations, and an expression blendshape to model facial expressions. The model can be expressed as $M(\boldsymbol{\beta}, \boldsymbol{\theta}, \boldsymbol{\psi}) : \mathbb{R}^{|\boldsymbol{\beta}| \times |\boldsymbol{\theta}| \times |\boldsymbol{\psi}|} \to \mathbb{R}^{3N}$, which generates $N$ vertices to represent a complete 3D face model using the shape parameters $\boldsymbol{\beta} \in \mathbb{R}^{|\boldsymbol{\beta}|}$, pose parameters $\boldsymbol{\theta} \in \mathbb{R}^{|\boldsymbol{\theta}|}$, and expression parameters $\boldsymbol{\psi} \in \mathbb{R}^{|\boldsymbol{\psi}|}$.

To generate all the vertices of a 3D face model for facial expression reconstruction, *mm3DFace* first retrieves the shape, pose, and expression parameters, denoted as $\beta$, $\theta$, and $\psi$, respectively. Figure 3.11 illustrates the procedure for obtaining the optimized parameters to create 3D facial avatars. Since *mm3DFace* primarily focuses on the continuous reconstruction of facial expressions, rather than the unique facial shape of an individual, we use fixed shape parameters $\hat{\beta}$ to represent a generalized face model.

Due to the high dimensionality and adaptability of the expression space, the pose parameters should be obtained prior to the expression parameters in order

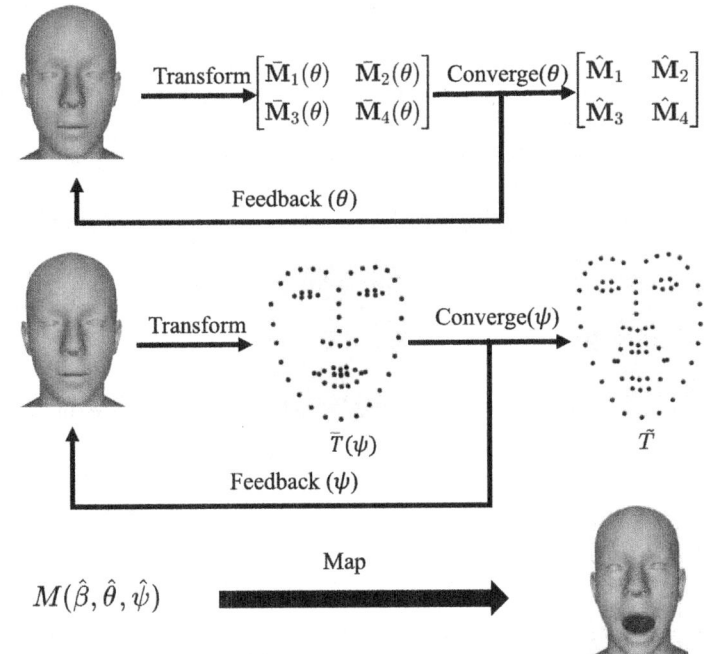

**Fig. 3.11** Illustration of FLAME-based 3D facial avatar generation

to prevent expression overfitting. *mm3DFace* first computes the pose parameters $\theta$ by utilizing the four affine matrices from different facial regions. Specifically, the pose parameters $\theta$ are optimized by minimizing the distance between the four affine matrices $\bar{\mathbf{M}}_i(\theta)$ derived from a 3D universal face model and the corresponding affine matrices $\hat{\mathbf{M}}_i$ from the facial shapes. In other words, $\bar{\mathbf{M}}_i(\theta)$ converges to $\hat{\mathbf{M}}_i$ as $\theta$ is iteratively adjusted, i.e.,

$$\hat{\theta} = \arg\min_{\theta} \sum_{i=1}^{4} \left\| \bar{\mathbf{M}}_i(\theta) - \hat{\mathbf{M}}_i \right\|_2^2, \tag{3.17}$$

where $\hat{\theta}$ represents the optimized pose parameters, $\bar{\mathbf{M}}_i(\theta)$ is the affine matrix for the $i_{th}$ region from the 3D face model using $\theta$, and $\hat{\mathbf{M}}_i$ is the affine matrix for the $i_{th}$ region from the facial shape data.

Next, *mm3DFace* retrieves the expression parameters using the 68 facial landmarks. Specifically, the expression parameters $\psi$ are optimized to minimize the discrepancy between the projected landmarks $\bar{T}(\psi)$ from the 3D face model's vertices and the landmarks $\tilde{T}$ derived from facial expressions. In other words, $\bar{T}(\psi)$ is iteratively adjusted to match $\tilde{T}$, i.e.,

$$\hat{\psi} = \arg\min_{\psi} \left\| \bar{T}(\psi) - \tilde{T} \right\|_2^2, \tag{3.18}$$

where $\hat{\psi}$ represents the optimized expression parameters, $\bar{T}(\psi)$ refers to the 68 landmarks generated by the 3D face model using $\psi$, and $\tilde{T}$ denotes the 68 landmarks obtained from the facial expressions.

Using the optimized shape parameters $\hat{\beta}$, pose parameters $\hat{\theta}$, and expression parameters $\hat{\psi}$, *mm3DFace* projects these parameters onto the vertices of the 3D face model [21] to create 3D facial avatars that dynamically express facial expressions.

## 3.6 Evaluation

In this section, we perform experiments to evaluate *mm3DFace* in real-world scenarios.

### 3.6.1 Evaluation Setup

We implement *mm3DFace* using a COTS mmWave radar TI AWR1443 [22]) connected to a data capture card (TI DCA1000EVM [23]). The radar system is equipped with three transmit antennas and four receive antennas. It is configured to emit mmWave chirp signals within a frequency range of 77–81 GHz, with a signal

## 3.6 Evaluation

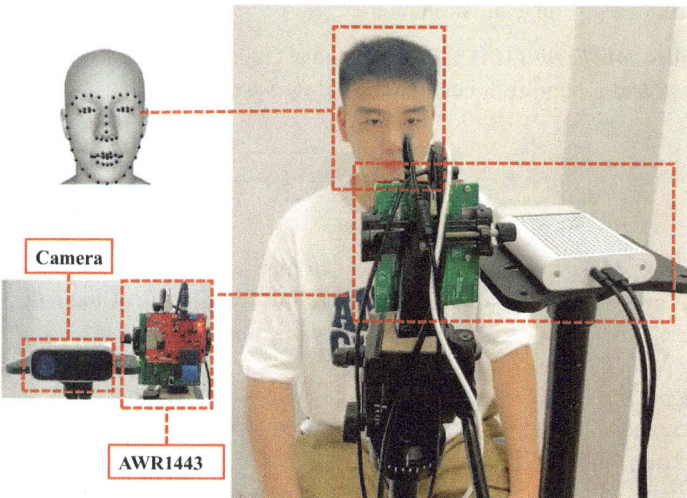

**Fig. 3.12** Experimental settings of *mm3DFace*

frame consisting of 128 pulses over a duration of 50 ms. The sampling rate is set to 512 points per pulse. For signal reading and processing, we use a PORSER W-2150B desktop. Additionally, we employ a depth camera to record human facial expressions and utilize Dlib [19] to generate the ground truth 68 facial landmarks in both 2D and 3D.

The experiments are carried out in three environments, including a laboratory, a meeting room, and a corridor. Figure 3.12 illustrates the experimental setup, where a mmWave radar continuously emits FMCW signals and records the reflected signals. A depth camera is positioned at the same location as the mmWave radar to capture depth images for ground truth. 15 volunteers take part in the experiments. The participants naturally and continuously engage in facial activities, such as facial movements, expressions (e.g., neutral, happy, sad, anger, surprise, fear, disgust, and contempt), and other facial behaviors at varying distances and orientations relative to the radar. The mmWave radar and depth camera continuously capture these facial activities.

We perform fivefold cross-validation, where distinct groups of users are assigned to provide training and testing data, respectively. Specifically, the dataset consisting of 15 users is divided into 5 subsets, with each subset containing data from 3 users. In each round of cross-validation, 80% of the data from 4 subsets is used for training, while the remaining 20% from those subsets, along with an additional subset, is utilized for testing and evaluation. The final experimental results are the average across all cross-validation rounds, which demonstrates *mm3DFace*'s performance on both seen and unseen users. For training, the learning rate is set to 0.0005, the batch size is 32, and the number of epochs is 500.

We introduce several evaluation metrics:

- *Mean Absolute Error (MAE)* measures the average error in distance between the predicted and groundtruth landmarks, and is expressed as:

$$MAE = \frac{1}{L}\sum_{i=1}^{L} \gamma\sqrt{(x_i - \bar{x}_i)^2 + (y_i - \bar{y}_i)^2},$$

where $L$ represents the total number of landmarks, $(x_i, y_i)$ are the coordinates of the $i$-th groundtruth landmark, $(\bar{x}_i, \bar{y}_i)$ are the coordinates of the $i$-th predicted landmark, and $\gamma$ is a scaling factor that converts pixel distances to millimeter distances.

- *3D-Mean Absolute Error (3D-MAE)* calculates the average error in distance between the 3D landmarks generated by the *mm3DFace* avatars and the 3D groundtruth landmarks, given by:

$$3D - MAE = \frac{1}{L}\sum_{i=1}^{L} \gamma\sqrt{(x_i - \bar{x}_i)^2 + (y_i - \bar{y}_i)^2 + (z_i - \bar{z}_i)^2}.$$

- *Normalized Mean Error (NME)* is the average error between the predicted and groundtruth landmarks, normalized by the interocular distance, and is given by:

$$NME = \frac{1}{LD}\sum_{i=1}^{L} \sqrt{(x_i - \bar{x}_i)^2 + (y_i - \bar{y}_i)^2},$$

where $D$ denotes the groundtruth distance between the left and right pupils.

- *F-score* is the harmonic mean of precision and recall, defined as:

$$F\text{-}score = 2 \cdot \frac{\text{precision} \cdot \text{recall}}{\text{precision} + \text{recall}}.$$

### 3.6.2 Overall Performance

We assess the overall performance of *mm3DFace* in reconstructing 3D human faces with varying facial expressions. As shown in Fig. 3.13, we present examples of 3D facial reconstructions for different expressions, including real-scene images, the generated 68 facial landmarks, and the corresponding 3D facial avatars. We can see that the 68 landmarks precisely capture the position and structure of facial features, clearly illustrating details of facial expressions including the motion of the eyes and mouth. Furthermore, the generated 3D facial avatars provide accurate animations

## 3.6 Evaluation

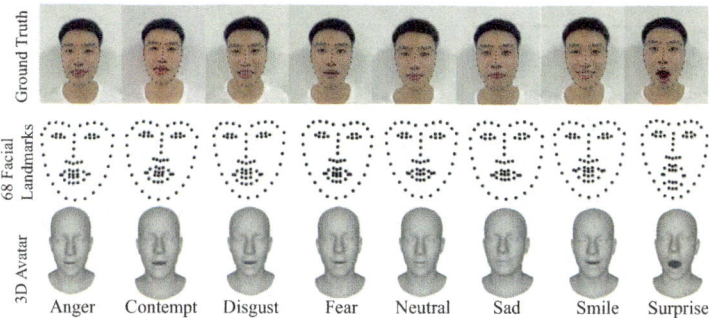

**Fig. 3.13** 3D facial reconstruction for different expressions

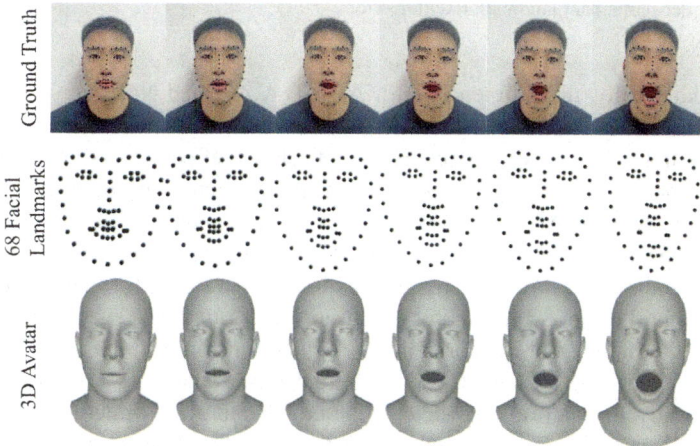

**Fig. 3.14** Facial reconstruction of continuous facial activity

that reflect the user's facial movements. These demonstrate that *mm3DFace* is effective in reconstructing 3D human faces with a wide range of facial expressions.

*mm3DFace* is also capable of reconstructing continuous facial movements and capturing subtle facial variations. In Fig. 3.14, we present an example of 3D facial reconstruction for a continuous and smooth facial activity. Additionally, this example highlights the system's ability to handle subtle facial changes, such as the gradual widening of the mouth, showcasing its proficiency in detecting and reconstructing delicate facial expressions.

We perform a quantitative evaluation of *mm3DFace* by computing the MAE, 3D-MAE, and NME for the 68 facial landmarks. As shown in Fig. 3.15, we report the overall MAE, 3D-MAE, and NME values for each user. On average, *mm3DFace* achieves an MAE of 2.30 mm, a 3D-MAE of 4.10 mm, and an NME of 3.94%, with standard deviations of 0.40, 0.63 mm, and 0.7%, respectively. These results demonstrate that *mm3DFace* provides accurate landmark tracking with minimal

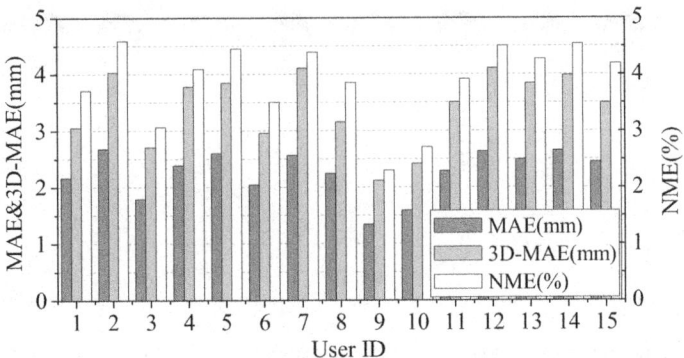

**Fig. 3.15** Per-participant landmark tracking error

error. Additionally, the differences in MAE, 3D-MAE, and NME across users are negligible. For instance, the discrepancies between user 2 and user 9 are only 1.34, 2.38 mm, and 2.29%, respectively. This confirms the effectiveness and robustness of *mm3DFace* in performing 3D facial reconstruction across different individuals.

We evaluate the MAE and 3D-MAE performance of *mm3DFace* across different facial regions. Figure 3.16a illustrates the CDF of MAE for the entire face as well as specific facial regions. From the figure, we observe that the performance in tracking facial landmarks shows only minor variations across regions. The MAE and 3D-MAE for the mouth and eyes are slightly higher compared to the nose and facial contour. This can be attributed to the fact that facial expressions, such as talking and other movements, primarily involve the mouth and eyes, while the contour and nose tend to remain relatively stable. Despite the slightly higher error rates for the mouth and eyes, the MAE and 3D-MAE values remain low overall. For example, 80% of the MAE and 3D-MAE values for the mouth are below 3.48 and 5.91 mm, respectively. Similarly, 80% of the MAE and 3D-MAE values for the eyes are below 2.96 and 5.91 mm, demonstrating that *mm3DFace* can still accurately capture the details of facial expressions in practical applications.

We also evaluate the performance for trained and untrained users. In the fivefold cross-validation process, four sub-datasets are used for both training and testing, while the remaining sub-dataset is used for testing. This setup allows us to assess the system's performance on both seen and unseen users. Figure 3.17 presents the 3D-MAE values for both trained and untrained users across each round of cross-validation. We can see that the difference in 3D-MAE between trained and untrained users is small, with an average difference of 0.08 mm and a maximum difference of 0.10 mm. Additionally, the variation between different rounds of cross-validation is negligible. These results show the robust generalization capabilities of *mm3DFace* when applied to new users.

## 3.6 Evaluation

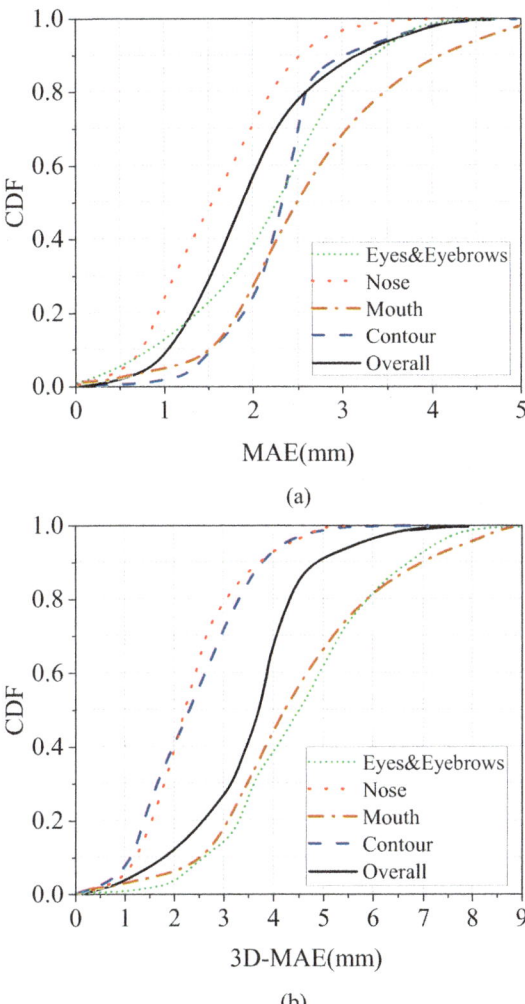

**Fig. 3.16** CDF of MAE and 3D-MAE of each facial region. (**a**) CDF of MAE. (**b**) CDF of 3D-MAE

### 3.6.3 Comparison with Existing Method

We compare the performance of facial landmark tracking using the proposed system with four existing methods: TCDCN [24], SAN [25], PFLD [26], and BioFace-3D [7]. Among these, TCDCN, SAN, and PFLD are vision-based techniques, while BioFace-3D are wearable sensor-based approach. We implement the vision-based methods and utilize public image datasets (i.e., 300-W [17] and IBUG [17]) for evaluation, measuring NME by comparing the predicted facial landmarks from images with the labeled groundtruth. BioFace-3D uses a custom-designed wearable device for facial landmark tracking. Since it is not feasible to replicate

**Fig. 3.17** 3D-MAE of trained/untrained users

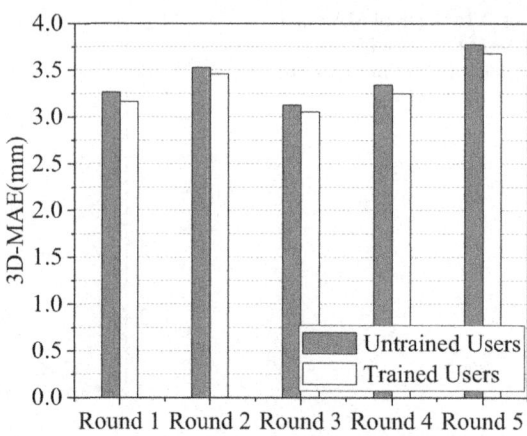

**Table 3.1** NME of *mm3DFace* and existing methods

| Methods | Dataset | Landmarks | NME |
|---|---|---|---|
| TCDCN [24] | 300-W | 68 | 5.54 |
| | IBUG | 68 | 8.60 |
| SAN [25] | 300-W | 68 | 3.98 |
| | IBUG | 68 | 6.60 |
| PFLD [26] | 300-W | 68 | 3.37 |
| | IBUG | 68 | 4.98 |
| BioFace-3D [7] | Self-collected | 53 | 3.38 |
| **mm3DFace** | **Self-collected** | **68** | **3.94** |

The bold values in the table are achieved by *mm3DFace*

the wearable device entirely, we use the reported results from the original paper [7] for comparison.

Table 3.1 presents the NME values for *mm3DFace* and four other methods that track different numbers of facial landmarks. As shown, *mm3DFace* achieves 3.94 NME with 68 facial landmarks, which is of similar performance with the vision-based methods. This demonstrates the potential of mmWave signal-based approaches to achieve effective 3D facial expression reconstruction comparable to vision-based techniques. In contrast the wearable sensor-based approach BioFace-3D demonstrates lower error rates, it only tracks 53 landmarks, resulting in a loss of detail in the facial contours. Thus, *mm3DFace* achieves comparable performance to existing methods.

### 3.6.4 Facial Expression Recognition

*mm3DFace* is also capable of recognizing various facial expressions. To enable recognition, we append a linear layer followed by a softmax layer after the 68 facial

3.6 Evaluation

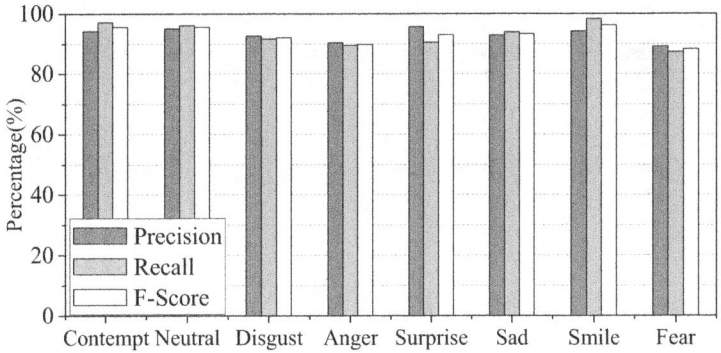

**Fig. 3.18** Performance of facial expression recognition

landmarks are predicted, and use CrossEntropy as the loss function in *mm3DFace*. We initialize the model with a pre-trained version that generates the 68 facial landmarks, freezing all parameters except the newly added linear layer for fine-tuning. In the experiment, users perform eight distinct facial expressions: contempt, neutral, disgust, anger, surprise, sadness, smile, and fear. To ensure precise labeling of these expressions, we continuously gather both mmWave signals and vision data from the camera throughout the evaluation period.

Figure 3.18 shows the precision, recall and F-score in recognizing 8 facial expressions. We can see that *mm3DFace* achieves an overall precision of 93.04%, recall of 93.06%, and F-score of 93.03% in facial expression recognition. The result demonstrates the effectiveness of *mm3DFace* in facial expression recognition. Besides, there are differences in recognition accuracy of different facial expressions. For example, the facial expressions of anger and fear have relatively lower precision because these facial expressions produce inapparent activities, making them difficult to distinguish.

### 3.6.5 Impact of Mask

To evaluate the performance of *mm3DFace* in NLOS conditions, we conduct experiments of reconstructing facial expressions when users wear masks. The evaluation involves two common types of masks: surgical masks and N95 masks. Figure 3.19 presents examples of 3D facial reconstruction for a user wearing masks while performing various facial expressions. The results demonstrate that *mm3DFace* effectively reconstructs the 3D face of the user even with masks.

Next, we perform a quantitative analysis to evaluate the effect of masks. In the experiment, mmWave signals are recorded while users wear masks, and the groundtruth images are captured after the masks are removed, ensuring that the users' facial expressions remain consistent throughout the process. Figure 3.20

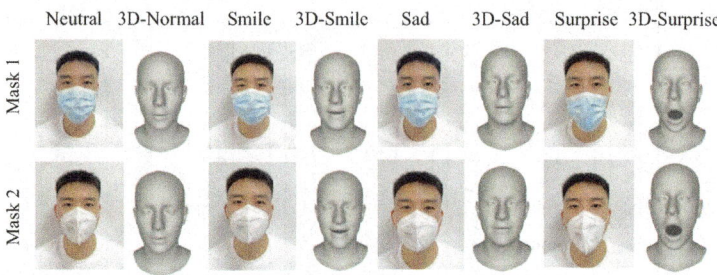

**Fig. 3.19** 3D facial reconstruction with masks

**Fig. 3.20** MAE and 3D-MAE with/without masks

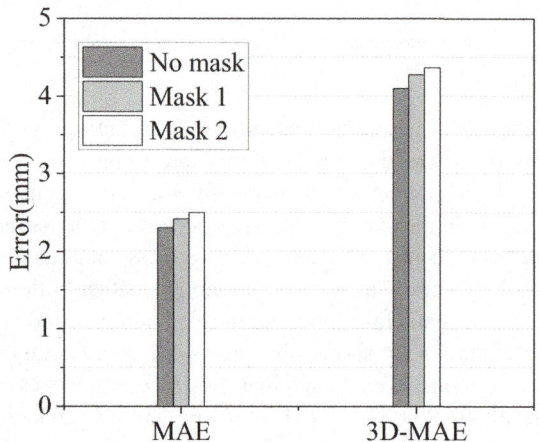

illustrates the MAE and 3D-MAE values for facial reconstruction with and without masks. The results indicate that the discrepancies between errors with and without masks are negligible. This can be attributed to the ability of mmWave signals to penetrate certain obstacles, effectively capturing the shape and movement of the face even when obscured by a mask. These findings show the robustness of *mm3DFace* in NLOS scenarios, demonstrating its potential for applications where traditional vision-based solutions are not feasible.

### 3.6.6 Impact of Distance

We evaluate the impact of distance between the face and the mmWave radar. The distance between the human face and the radar ranges from 20 to 120 cm, and we manually annotate the collected mmWave data by measuring the distance for evaluation purposes. Figure 3.21 presents the MAE and 3D-MAE for different facial regions at various distances. When the distance is between 20 and 60 cm, both the MAE and 3D-MAE for different facial regions remain relatively constant and show

## 3.6 Evaluation

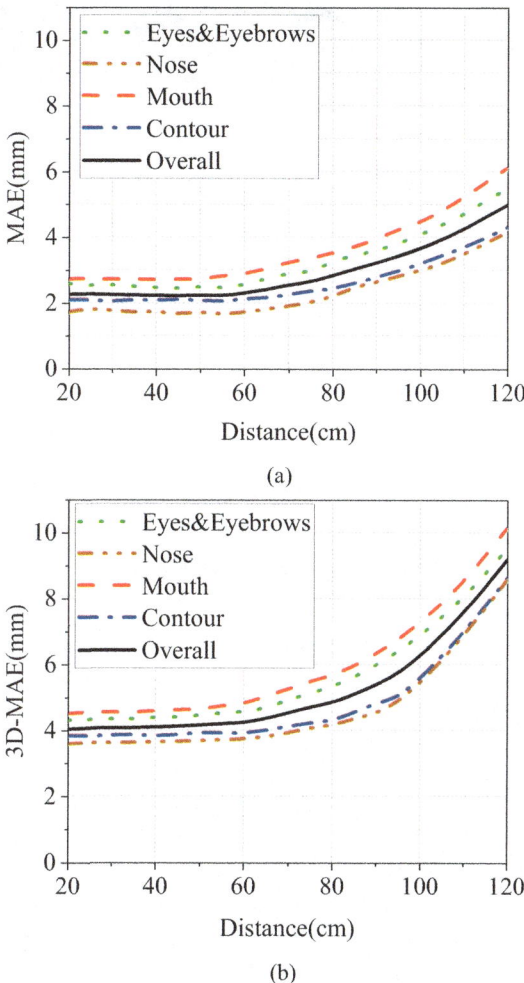

**Fig. 3.21** MAE and 3D-MAE in different distances. (**a**) MAE. (**b**) 3D-MAE

low error levels. For example, at a distance of 20 cm, the MAE and 3D-MAE for the entire face are 2.30 and 4.11 mm, respectively, while at 60 cm, these values are 2.34 and 4.24 mm. However, as the distance increases from 60 to 120 cm, both the MAE and 3D-MAE show a gradual increase due to the attenuation of the mmWave signals over greater distances, which makes it more challenging to capture facial features effectively. Despite this, even at a distance of 120 cm, the MAE and 3D-MAE are 5.01 and 9.21 mm respectively, which is still reasonable expression recognition performance in practice.

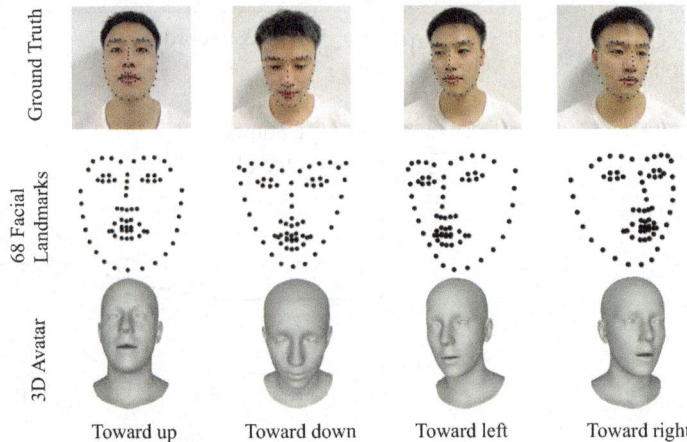

**Fig. 3.22** 3D facial reconstruction in different orientations

## 3.6.7 Impact of Orientation

We evaluate the performance of *mm3DFace* when the human face is oriented at different angles relative to the mmWave radar. In this experiment, users' faces are positioned in four directions (i.e., up, down, left, and right) at angles ranging from 0° to 45° with 15° increments, at a distance of 60 cm. Figure 3.22 shows examples of 3D facial reconstruction at a 30° angle in each of the four directions. As shown in the figure, the 68 facial landmarks are rotated to specific angles depending on the face's orientation relative to the radar. The 3D avatars faithfully reconstruct the user's facial expressions across these various orientations.

We then perform a quantitative evaluation of the MAE and 3D-MAE across different facial orientations. Figure 3.23 illustrates the MAE and 3D-MAE values from 0° to 45°, with 15° intervals, for the four directions. The results show that there is minimal variation in the facial landmark tracking errors across different orientations. While the tracking errors slightly increase at 45°, the average MAE and 3D-MAE are still low, at 2.50 and 4.45 mm, respectively, which are sufficient for accurate reconstruction of facial expressions. These results demonstrate that the affine transformation efficiently rotates the facial landmarks to match the true orientation of the face.

## 3.6.8 Impact of Background Environment

We evaluate the performance of *mm3DFace* in various background settings, such as an empty lab, a meeting room with intricate layouts, and a corridor. The experiments are carried out both during the day and at night to test the system's robustness

**Fig. 3.23** MAE and 3D-MAE in different orientations. (**a**) MAE. (**b**) 3D-MAE

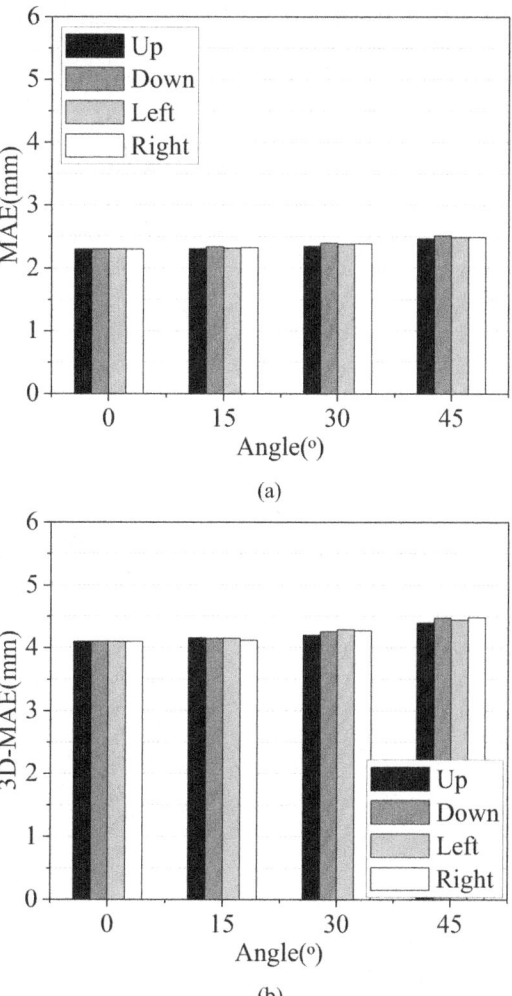

under different lighting conditions. For the nighttime analysis, mmWave signals are recorded in total darkness, while the groundtruth images are captured with additional lighting at the same time, ensuring the users' facial expressions remain consistent. Figure 3.24 illustrates the MAE and 3D-MAE for facial landmark tracking across different environments during both day and night. As shown, *mm3DFace* demonstrates strong resilience to varying lighting conditions, even in complete darkness. Furthermore, the differences in MAE and 3D-MAE across different environments are small. While the MAE and 3D-MAE in the corridor are slightly higher compared to the lab and meeting room, likely due to interference from passers-by affecting the mmWave signals, the differences are small, such as 0.09 and 0.20 mm between the corridor and lab. This consistency is attributed to the

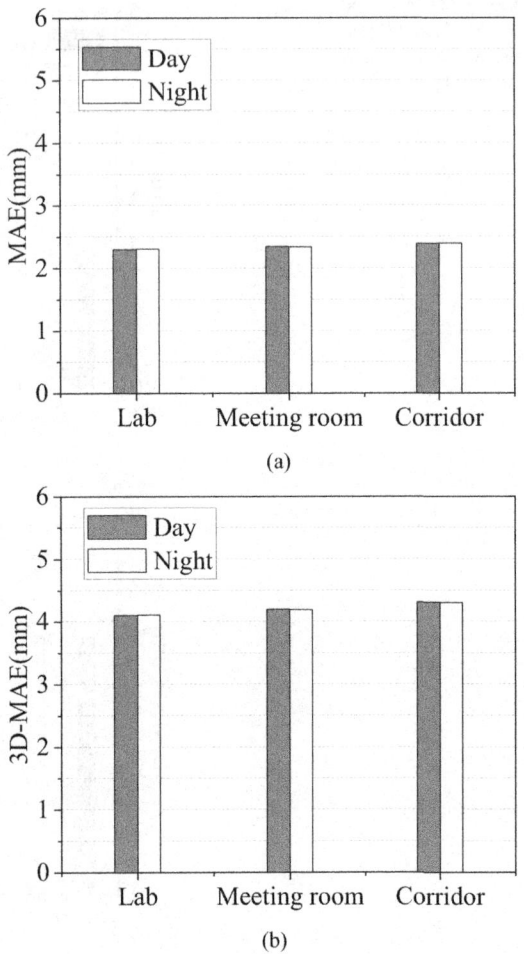

**Fig. 3.24** MAE and 3D-MAE in 3 environments. (**a**) MAE. (**b**) 3D-MAE

effectiveness of the ConNeXt model, which can accurately detect the spatial position of the face and capture its features from mmWave signals even with background changes.

## 3.6.9 Time Consumption

We evaluate the time efficiency of *mm3DFace* to measure its real-time performance in reconstructing 3D facial expressions. Time consumption is defined as the delay between the moment a mmWave sample is received and the time the system produces the corresponding facial expressions. *mm3DFace* operates on a desktop featuring an Intel i9-12900K processor and an NVIDIA RTX3090Ti GPU as the

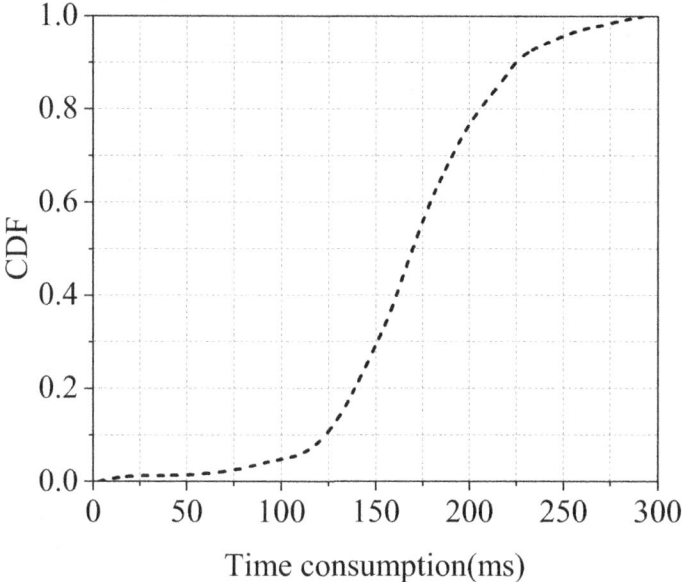

**Fig. 3.25** CDF of time consumption

backend. Figure 3.25 presents the CDF of *mm3DFace*'s time consumption. The results reveal that the average processing time is 178 ms, with 90% of the time instances falling below 225 ms. This demonstrates that the system can generate the reconstructed facial expressions with minimal delay, ensuring a smooth and responsive user experience.

## 3.7 Summary

In this chapter, we introduce *mm3DFace*, a nonintrusive 3D facial expression reconstruction system that utilizes a COTS mmWave radar to track facial landmarks and reconstruct 3D facial expressions. It first employs a ConvNeXt model with triple loss embedding to extract geometric features of the face. Then, a region-divided affine transformation is designed to reconstruct distance- and orientation-robust facial shapes. Afterward, a regional amplification method is applied to reconstruct detailed facial expressions, culminating in the generation of continuous facial avatars. Extensive experiments conducted in real-world environments validate the effectiveness of *mm3DFace* in achieving accurate 3D facial expression reconstruction using mmWave radar.

# References

1. Poria, S., Cambria, E., Bajpai, R., Hussain, A.: A review of affective computing: From unimodal analysis to multimodal fusion. Inf. Fusion **37**, 98–125 (2017)
2. Kemelmacher-Shlizerman, I., Basri, R.: 3d face reconstruction from a single image using a single reference face shape. IEEE Trans. Pattern Anal. Mach. Intell. **33**(2), 394–405 (2010)
3. Richardson, E., Sela, M., Or-El, R., Kimmel, R.: Learning detailed face reconstruction from a single image. In: Proceedings of the IEEE CVPR '17. Honolulu (2017)
4. Jiang, L., Zhang, J., Deng, B., Li, H., Liu, L.: 3d face reconstruction with geometry details from a single image. IEEE Trans. Image Process. **27**(10), 4756–4770 (2018)
5. Matthies, D.J., Strecker, B.A., Urban, B.: Earfieldsensing: a novel in-ear electric field sensing to enrich wearable gesture input through facial expressions. In: Proceedings of the ACM CHI '17. Denver (2017)
6. Nam, Y., Koo, B., Cichocki, A., Choi, S.: Gom-face: Gkp, eog, and emg-based multimodal interface with application to humanoid robot control. IEEE Trans. Biomed. Eng. **61**(2), 453–462 (2013)
7. Wu, Y., Kakaraparthi, V., Li, Z., Pham, T., Liu, J., Nguyen, P.: Bioface-3d: continuous 3d facial reconstruction through lightweight single-ear biosensors. In: Proceedings of the ACM MobiCom '21. New Orleans (2021)
8. Wu, C., Zhang, F., Wang, B., Liu, K.R.: mmtrack: Passive multi-person localization using commodity millimeter wave radio. In: Proceedings of the IEEE INFOCOM '20. Toronto (2020)
9. Cui, H., Dahnoun, N.: High precision human detection and tracking using millimeter-wave radars. IEEE Aerospace Electron. Syst. Mag. **36**(1), 22–32 (2021)
10. Liu, H., Wang, Y., Zhou, A., He, H., Wang, W., Wang, K., Pan, P., Lu, Y., Liu, L., Ma, H.: Real-time arm gesture recognition in smart home scenarios via millimeter wave sensing. Proc. ACM IMWUT' 20 **4**(4), 1–28 (2020)
11. Zhang, R., Cao, S.: Real-time human motion behavior detection via cnn using mmwave radar. IEEE Sensors Lett. **3**(2), 1–4 (2018)
12. Xue, H., Ju, Y., Miao, C., Wang, Y., Wang, S., Zhang, A., Su, L.: mmmesh: towards 3d real-time dynamic human mesh construction using millimeter-wave. In: Proceedings of the ACM MobiSys '21. Wisconsin (2021)
13. Kong, H., Xu, X., Yu, J., Chen, Q., Ma, C., Chen, Y., Chen, Y.C., Kong, L.: m3track: mmwave-based multi-user 3d posture tracking. In: Proceedings of the ACM MobiSys' 22. Portland (2022)
14. Xue, H., Cao, Q., Ju, Y., Hu, H., Wang, H., Zhang, A., Su, L.: M4esh: mmwave-based 3d human mesh construction for multiple subjects. In: Proceedings of the ACM SenSys '22, pp. 391–406. Boston (2022)
15. Liu, Z., Mao, H., Wu, C.Y., Feichtenhofer, C., Darrell, T., Xie, S.: A convnet for the 2020s. In: Proceedings of the IEEE/CVF CVPR' 22. New Orleans (2022)
16. Weinberger, K.Q., Saul, L.K.: Distance metric learning for large margin nearest neighbor classification. J. Mach. Learn. Res. **10**(2), 207–241 (2009)
17. Sagonas, C., Antonakos, E., Tzimiropoulos, G., Zafeiriou, S., Pantic, M.: 300 faces in-the-wild challenge: database and results. Image Vision Comput. **47**, 3–18 (2016)
18. Weisstein, E.W.: Affine Transformation (2004). https://mathworld.wolfram.com/
19. King, D.E.: A toolkit for making real world machine learning and data analysis applications in c++ (2022) [Online]. Available: https://github.com/davisking/dlib
20. Girshick, R.: Fast r-cnn. In: Proceedings of the of IEEE ICCV '15. Santiago (2015)
21. Li, T., Bolkart, T., Black, M.J., Li, H., Romero, J.: Learning a model of facial shape and expression from 4d scans. ACM Trans. Graph. **36**(6), 194–205 (2017)
22. Texas Instruments: Awr1443 single-chip 76-ghz to 81-ghz automotive radar sensor integrating mcu and hardware accelerator (2021) [Online]. Available: https://www.ti.com/product/AWR1443

# References

23. Texas Instruments: Dca1000evm real-time data-capture adapter for radar sensing evaluation module (2021) [Online]. Available: https://www.ti.com/tool/DCA1000EVM
24. Zhang, Z., Luo, P., Loy, C.C., Tang, X.: Facial landmark detection by deep multi-task learning. In: Proceedings of the Springer ECCV '14. Zurich (2014)
25. Dong, X., Yan, Y., Ouyang, W., Yang, Y.: Style aggregated network for facial landmark detection. In: Proceedings of the CVF/IEEE CVPR '18. Salt Lake City (2018)
26. Guo, X., Li, S., Yu, J., Zhang, J., Ma, J., Ma, L., Liu, W., Ling, H.: Pfld: A practical facial landmark detector. Preprint (2019). arXiv:1902.10859

# Chapter 4
# mmWave-based Hand Gesture Reconstruction

**Abstract** Hand gesture reconstruction plays a crucial role in enabling a range of interactive technologies such as human-computer interaction, sign language interpretation, virtual reality simulation, and more. Current methodologies predominantly rely on wearable devices like gloves or wristbands to capture hand gestures, which often involves high deployment costs and can result in a disruptive user experience. Alternatively, some techniques leverage vision-based systems, but these can suffer from challenges of varying lighting conditions and potential privacy concerns. In this chapter, we introduce *mmHand*, a novel system for 3D hand pose reconstruction based on mmWave radar signals. This approach enables the generation of 3D hand skeletons and the reconstruction of 3D hand meshes. First, *mmHand* uses mmWave signals to detect the hand and process the acquired mmWave data. Then, it applies a custom-designed attention-driven hourglass network, *mmSpaceNet*, to capture spatial features and employs LSTM networks to extract temporal characteristics. With the extracted features, *mmHand* then performs a regression to predict hand joint locations in 3D space, generating accurate 3D hand skeletons. Lastly, 3D hand meshes that provide a detailed and continuous representation of hand poses are reconstructed through a hand Model with Articulated and Non-rigid defOrmations (MANO). Extensive experimental evaluations show that *mmHand* achieves a mean per-joint position error of 18.3 mm and 95.1% accuracy, demonstrating its effectiveness in hand pose estimation.

**Keywords** Millimeter wave · Hand pose estimation · Hand joint regression · Hand mesh reconstruction

## 4.1 Introduction

Human hands, as a vital means of expression, convey a wide range of personal desires, intentions, and actions through various gestures. Therefore, hand-based interaction serves as one of the most intuitive and ubiquitous methods to bridge the gap between humans and machines. Hand gesture reconstruction refers to the technique of continuously estimating and modeling hand gestures and movements, which forms the core knowledge that enables machines to interpret interactive content. With the rapid growing of IoT applications, hand pose estimation has become an essential component in numerous interaction contexts, including user interface control, sign language recognition, VR, AR, interactive gaming, and more.

To address the demand for interaction-driven applications, both academic researchers and industry professionals have developed different approaches for hand gesture reconstruction. Among these, wearable-based methods are dominant technologies that have gained widespread adoption. Devices like data gloves [1] and wristbands [2] are capable of capturing highly precise hand shapes and movements by employing sensitive sensors attached to the hand, making them ideal solutions for scenarios such as medical surgeries. However, wearable devices are typically designed for specific tasks and come with high setup costs. Moreover, the need to actively wear these devices can lead to a cumbersome user experience and restrict usage scenarios. In contrast to wearable solutions, using images or videos for hand pose estimation has obtained considerable attention. Vision-based methods often rely on large-scale datasets and deep neural networks to reconstruct hand skeletons or generate hand meshes [3–7]. However, these approaches are highly sensitive to lighting conditions and struggle in non-line-of-sight environments. Additionally, vision-based techniques may expose personal information from the background, which could raise privacy concerns. As a result, there is a strong demand for a cost-effective, passive, and non-intrusive solution for hand gesture reconstruction.

In recent years, radio frequency signals have found increasing use in sensing applications beyond communications, with mmWave signals emerging as one of the most prominent choices. While mmWave signals have demonstrated success in gesture recognition [8–10], these methods are typically limited to classifying predefined gestures and are unable to capture dynamic 3D hand poses. A recent study [11] employs mmWave signals to detect human forearms and infer finger movements. However, this approach overlooks the palm's shape and cannot produce realistic 3D hand meshes. Additionally, the method requires the forearm to remain aligned with the radar, which significantly degrades performance when users rotate their arms. Therefore, the increasing demand for accurate 3D hand reconstruction motivates the development of a nonintrusive, realistic 3D hand gesture reconstruction based on mmWave signals. Such a system is resilient to varying lighting conditions, operate in a privacy-preserving manner, and offer a nonintrusive solution that can be easily deployed for a wide range of interactive applications, such as user interface control, virtual reality modeling, etc. Achieving hand gesture reconstruction using mmWave signals presents several practical challenges. First, capturing the complex and nuanced movements of a hand requires the system to robustly track subtle motions using a COTS mmWave radar. Second, given the relatively low resolution and inherent errors of mmWave signals, it is crucial to effectively extract multi-scale features from these signals to accurately represent the human hand. Lastly, to enable real-world applications, the system must be capable of generating dynamic 3D hand skeletons and reconstructing realistic 3D hand meshes.

In this chapter, we introduce a nonintrusive 3D hand gesture reconstruction system, *mmHand*, which continuously generates 3D hand skeletons and reconstructs

## 4.1 Introduction

hand meshes using a mmWave radar. *mmHand* first utilizes mmWave signals to detect a user's hand and pre-process the data. Next, *mmHand* extracts spatial features through a specially designed attention-based hourglass network, *mmSpaceNet*, and captures temporal features via LSTM networks. Using the extracted features, *mmHand* performs 3D joint regression to generate 3D hand skeletons by combining 3D loss and kinetics loss functions. In the final step, *mmHand* reconstructs 3D hand meshes, which provide continuous and detailed representations of hand poses using the MANO. We evaluate the performance of *mmHand* through a series of experiments conducted in real-world environments. The results demonstrate that *mmHand* accurately estimates a variety of hand poses. A schematic of the *mmHand* system is presented in Fig. 4.1.

We summarize the key contributions as follows:

- We introduce *mmHand*, a nonintrusive 3D hand gesture reconstruction system that utilizes mmWave signals to produce 3D hand skeletons and continuously reconstruct 3D hand meshes.
- We develop a deep learning architecture that effectively extracts multi-scale spatial and temporal features from mmWave signals for accurate hand gesture reconstruction.
- We perform 3D joint regression to estimate 21 hand joints, generating dynamic hand skeletons and subsequently reconstructing realistic and continuous 3D hand meshes.
- We conduct extensive experiments with 10 participants in real-world settings. The results demonstrate that *mmHand* accurately reconstructs 3D hand meshes, achieving a mean per joint position error of 18.3 mm and 95.1% correct keypoints for 3D hand joint estimation.

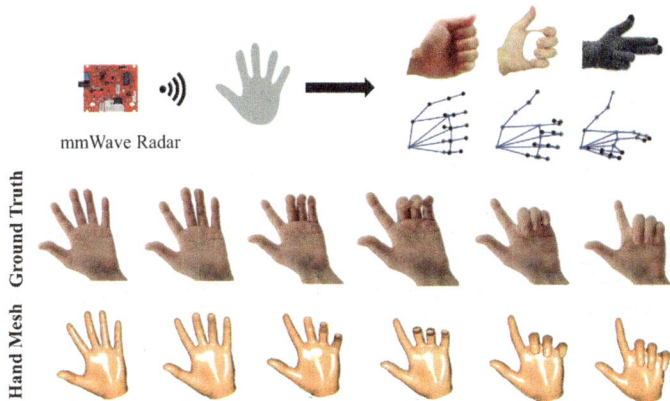

**Fig. 4.1** Illustration of *mmHand*, which utilizes a mmWave radar to realize 3D hand pose estimation

## 4.2 System Overview

To enable hand gesture reconstruction in real-world environments, we develop *mmHand*, which utilizes a commercial mmWave radar to continuously capture the motion and shape of hands for hand gesture reconstruction. Figure 4.2 illustrates the architecture of *mmHand*, which comprises the following modules.

**mmWave Signal Pre-processing** In this module, *mmHand* utilizes mmWave signals to detect the user's hand and preprocess the data. The mmWave radar captures signals reflected from the hand, and *mmHand* subsequently extracts distance, velocity, and angle information through a sequence of FFT operations. This preprocessing step offers vital information about the hand's posture and movement, forming the foundation for the subsequent hand gesture reconstruction.

**Hand Joint Regression** In this module, *mmHand* constructs 3D hand skeletons from the pre-processed mmWave signals using a specially designed deep learning model. First, *mmHand* extracts multi-scale spatial features that capture the hand's posture through an attention-based hourglass network, *mmSpaceNet*. Next, temporal features that represent the hand's motion are extracted using a LSTM-based temporal model. Finally, *mmHand* performs 3D joint regression and generates the 3D hand skeletons by combining the extracted features with both 3D loss and kinetics loss functions.

**Hand Mesh Reconstruction** In this module, *mmHand* further reconstructs 3D hand meshes by utilizing the universal parametric hand model MANO. By optimizing the pose and shape parameters of MANO using the regressed 3D hand joints, *mmHand* generates highly realistic 3D meshes of the hand, enabling dynamic hand gesture reconstruction with mmWave signals.

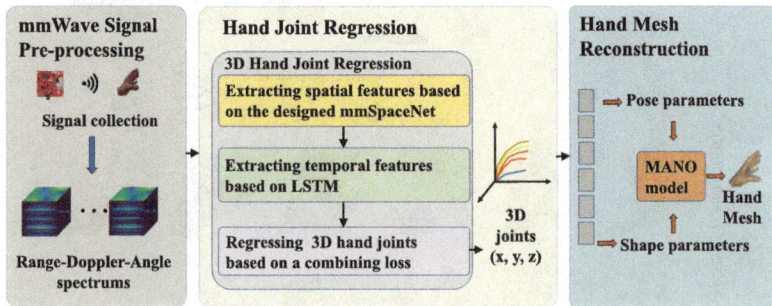

**Fig. 4.2** Architecture of *mmHand*

## 4.3 Signal Pre-Processing

To detect the posture and motion of a hand, *mmHand* first measures the range, velocity, and angle of the target. Specifically, a mmWave radar emits chirp signals with linearly increasing frequencies from the transmit antennas. These signals are reflected by objects in the environment and captured by the radar's receive antennas. The transmitted and received signals are then mixed in the radar's mixer, generating IF signals, which are represented as

$$x_{IF}(t) = A_r \cdot e^{j2\pi[f_0 + \frac{B}{T_c}t - \frac{B}{2T_c}\tau(r,c)]}, \tag{4.1}$$

where $f_0$ denotes the initial frequency of the chirp, $B$ is the signal bandwidth, $T_c$ is the duration of the chirp, $A_r$ is the amplitude factor accounting for the attenuation of mmWave signals, and $\tau(r, c)$ represents the time delay of the received signals relative to the transmitted ones. This delay is determined by the propagation speed $c$ of the mmWave signals and the distance $r$ between the object and the radar.

After obtain the IF signals, *mmHand* performs several signal processing operations to compute the range, velocity, and angle of the hand. The range $r$ between the radar and an object can be expressed as $r = \frac{cfT_c}{2B}$, where $f$ represents the frequency of the IF signals. However, mmWave signals could be affected by environmental noise, which impacts the accuracy of hand detection. Hence, it is essential to first mitigate environmental interference in the received mmWave signals. Since the range $r$ is directly related to the frequency $f$, different frequencies in the IF signals correspond to the hand and various objects in the environment, such as the human body, furniture, etc. As depicted in Fig. 4.3, the hand, human body, and furniture produce distinct peaks in the spectrum due to their varying distances from the radar, with the hand typically appearing as the first dominant peak, as it is generally closest to the radar during gesture interactions. To eliminate environmental disturbances, *mmHand* applies an 8th-order bandpass Butterworth filter to the raw mmWave signals, preserving only the components related to the hand. Finally, by performing range-FFT on the filtered mmWave signals, *Range-Spectrums* that represent the objects in the range dimension are extracted.

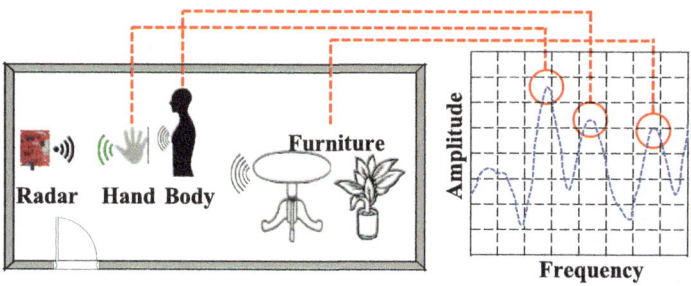

**Fig. 4.3** Illustration of sensing human hand and other objects using mmWave signals

To compute the velocity $v$ of an object, the FMCW radar transmits two chirps separated by a time interval of $T_c$. Upon performing range-FFT, both signals will peak at the same position in the *Range-Spectrums*, but they exhibit a phase shift $\Delta\phi$, which is proportional to the object's movement over the duration of $vT_c$. Hence, the velocity $v$ can be determined using the equation $v = \frac{\lambda \Delta\phi}{4\pi T_c}$, where $\lambda$ is the wavelength of the signals. After applying Doppler-FFT, the *Doppler-Spectrums* are derived.

To compute the AoA, a minimum of two receive antennas is necessary. The difference in distance $\Delta d$ between the object and the two antennas introduces a phase shift $\Delta\phi$ at the peak of the range-FFT, which can be expressed as $\Delta\phi = \frac{2\pi \Delta d}{\lambda}$. Based on geometric principles, $\Delta d$ is related to the AoA by $\Delta d = l \sin(\theta)$, where $l$ represents the distance between the two receive antennas, and $\theta$ is the AoA. Therefore, $\theta$ can be computed as $\theta = \sin^{-1}\left(\frac{\lambda \Delta\phi}{2\pi l}\right)$. To locate the object in space, *mmHand* employs TDM-MIMO technology to calculate two types of AoA, i.e., azimuth and elevation. The four receive antennas are continuously active, while the three transmit antennas are sequentially activated, creating virtual antenna arrays that allow simultaneous measurement of both azimuth and elevation. Next, *mmHand* applies angle-FFT on the signals to derive the *Azimuth-Spectrums* and *Elevation-Spectrums*. However, the angular resolution of the spectrum obtained through conventional fast Fourier transform is limited. Since the hand is typically located within a ±30° range relative to the radar's azimuth and elevation, *mmHand* utilizes zoom-FFT with a refinement factor of 2 in angle-FFT to enhance the accuracy of angle estimation. Finally, the *Azimuth-Spectrums* and *Elevation-Spectrums* are produced.

After signal preprocessing, *mmHand* forms a four-dimensional matrix that includes all spectrums, i.e., *Range-Spectrums*, *Doppler-Spectrums*, *Azimuth-Spectrum*, and *Elevation-Spectrums*, which consists of a *Radar Cube*. The *Radar Cube* encodes the range, angle, and velocity information of the detected hand.

## 4.4 Hand Joint Regression

To model a human hand, we utilize a well-known 21-joint hand model, consisting of a wrist joint, 16 finger joints, and 4 fingertip joints, as shown in Fig. 4.4. For mmWave-based hand gesture reconstruction, we introduce a deep neural network in *mmHand* that regresses the positions of the 21 hand joints in 3D space to generate 3D hand skeletons.

The input to the designed deep learning model is the *Radar Cube* (**RC**), which is derived from the pre-processed signals. The **RC** is represented as a four-dimensional matrix, denoted by $\mathbf{RC} \in \mathbb{R}^{F \times V \times D \times A}$, where $F$ refers to the number of frames, $V$ is the count of velocity bins, $D$ is the number of distance bins, and $A$ represents the number of angle bins. In theory, feeding each frame of **RC** into the neural network should yield 3D hand skeletons corresponding to that specific frame. However, the

## 4.4 Hand Joint Regression

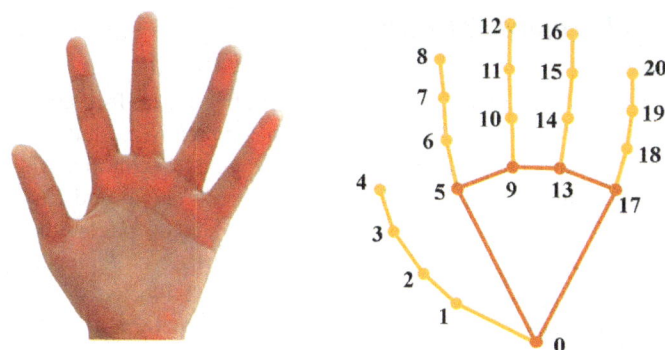

**Fig. 4.4** The 21-hand-joint model used in *mmHand*

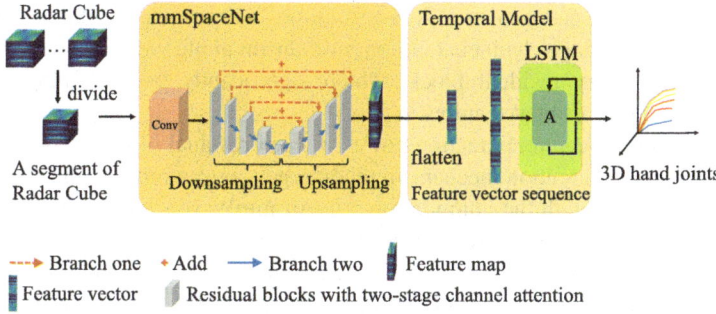

**Fig. 4.5** The architecture of 3D hand joint regression in *mmHand*

features extracted from a single frame fail to capture the full range of hand features at a given moment, resulting in inaccurate hand joint regression. Therefore, we use multiple consecutive frames to create a segment of the *Radar Cube*, denoted as $\mathbf{X} \in \mathbb{R}^{st \times V \times D \times A}$, where $st$ is the number of consecutive frames in each segment, and $\mathbf{X}$ is a single input to the network. The radar cube, encompassing more temporal information, offers a detailed representation of hand motions at a specific instance.

For effective hand pose estimation using mmWave signals, *mmHand* must capture the spatial arrangement of the hand in 3D space while also characterizing its motion over time. Therefore, *mmHand* first extracts spatial and temporal features of the hand from the mmWave signals and then regresses the positions of 21 hand joints in 3D space to construct 3D hand skeletons. The architecture of hand joint regression in *mmHand* is illustrated in Fig. 4.5.

### 4.4.1 Hand Feature Extraction

Given that a human hand has a relatively small reflective surface and its postures are highly varied, the reflected mmWave signals typically exhibit low reflection intensity and remain quite similar across different hand positions. To efficiently extract spatial features of the hand, we propose an attention-driven hourglass network, *mmSpaceNet*, which integrates both shallow and deep features to represent the human hand at various levels of spatial granularity. As depicted in Fig. 4.5, *mmSpaceNet* is composed of attention-based residual blocks, with each block featuring two branches. One branch modifies the channel count without altering the feature map size using 1 × 1 convolutional layers to preserve the current-level features. The second branch employs convolutional layers for downsampling, capturing high-dimensional and fine-grained features, followed by deconvolutional layers to upsample and generate high-resolution feature maps. Additionally, we incorporate a dual-stage channel attention mechanism along with a spatial attention mechanism in every residual block, which significantly boosts *mmSpaceNet*'s capability to extract critical features.

To improve the feature extraction power of each residual block, we introduce a dual-stage channel attention mechanism that merges conventional channel attention techniques [12, 13] with the unique properties of mmWave signals. An illustration of the two-stage attention mechanism is provided in Fig. 4.6.

Each segment of the *Radar Cube* $\mathbf{X}$ can be viewed as a sequence of $st$ three-dimensional matrices, where $\mathbf{X} = [X_1, X_2, \ldots, X_{st}] \in \mathbb{R}^{st \times V \times D \times A}$. For each $X_i \in \mathbb{R}^{V \times D \times A}$, $i = 1, 2, \ldots st$, the initial channel attention mechanism is applied, which is defined as:

$$a_i = \sigma(\text{Conv}_1(\text{TGAP}(X_i) + \text{TGMP}(X_i))), \tag{4.2}$$

$$Y_i = a_i X_i, \tag{4.3}$$

where $a_i$ denotes the attention weight for the $i$-th frame channel, $\sigma$ is the sigmoid activation function, $\text{Conv}_1$ indicates a block comprising two convolutional layers, TGAP refers to three-dimensional global average pooling, and TGMP refers to three-dimensional global max pooling. Following this, $X_i$ is scaled by the corresponding weight $a_i$, and the resulting output for each frame channel

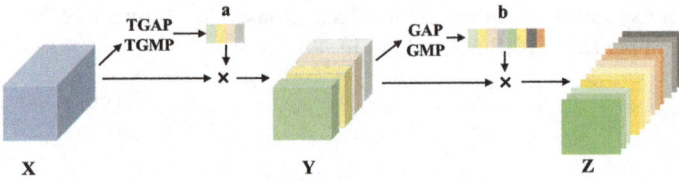

**Fig. 4.6** Two-Stage channel attention mechanism

## 4.4 Hand Joint Regression

is denoted as $Y_i$. After the first phase, the original input $\mathbf{X}$ is transformed into $\mathbf{Y} = [Y_1, Y_2, \ldots, Y_{st}] \in \mathbb{R}^{st \times V \times D \times A}$, with frame channel weighting applied.

Next, the second stage attention mechanism is applied. A global max pooling (GMP) and a global average pooling (GAP) are performed across each velocity channel. The outcomes of these two pooling operations are concatenated to form the channel features, ensuring that more information is preserved. Subsequently, a fully connected layer FC encodes all the channel features into a weight vector, which is then multiplied by the original input $\mathbf{Y}$. This process is mathematically described as:

$$b_i = \sigma(\text{FC}([\text{GAP}(Y_i), \text{GMP}(Y_i)])), \tag{4.4}$$

$$Z_i = b_i Y_i. \tag{4.5}$$

Subsequently, the input $\mathbf{Y}$ is transformed into $\mathbf{Z} = [Z_1, Z_2, \ldots, Z_{st}] \in \mathbb{R}^{st \times V \times D \times A}$, which is now weighted across the velocity channel.

The mmWave radar processes and receives reflected signals uniformly across all distances and directions. However, for hand gesture reconstruction, not all positions hold the same significance. We pay more attention to the finger joints and fingertips, which correspond to specific regions on the *Range-angle Spectrums*. To enable the network to distinguish between these regions in the *Range-angle Spectrums*, we apply a spatial attention mechanism on the output $\mathbf{Z}$, which is obtained after the two-stage channel attention mechanism. This is expressed as:

$$C_i = \sigma(\text{Conv}_2([\text{MEAN}(Z_i), \text{MAX}(Z_i)])), \tag{4.6}$$

$$W_i = C_i Z_i, \tag{4.7}$$

where MEAN refers to the mean of all feature maps of size $D \times A$ computed along the velocity dimension, MAX represents the maximum value computed across all feature maps along the velocity dimension, and $\text{Conv}_2$ is a convolutional layer that adjusts the number of channels. After applying the spatial attention mechanism, the input $\mathbf{Z}$ is transformed into $\mathbf{W} = [W_1, W_2, \ldots, W_{st}] \in \mathbb{R}^{st \times V \times D \times A}$. In comparison to the original input $\mathbf{X}$, the tensor $\mathbf{W}$ highlights the importance of different frame and velocity channels, placing greater focus on the critical regions of the *Range-angle Spectrums*. After the aforementioned processing, a feature map containing multi-scale spatial information of the hand is derived from the mmWave signals. This feature map is then fed into the temporal module to further capture the temporal dynamics of the hand.

To enable dynamic hand pose estimation, we introduce a temporal model aimed at extracting temporal features, as illustrated in Fig. 4.5. Since adjacent mmWave frames exhibit high correlation, we incorporate LSTM into the temporal model to capture the continuous motion of the hand. In particular, each segment of the *Radar Cube* $\mathbf{X}$ undergoes processing in *mmSpaceNet* to produce a global feature map. This global feature map is then flattened before being fed into the temporal model. Then, each individual network input results in the generation of a feature vector. All of these feature vectors are concatenated to form a sequence, which is fed into the LSTM to extract temporal features.

### 4.4.2 Regressing 3D Hand Joints Based on a Combined Loss

After extracting spatial and temporal features, *mmHand* further regresses 21 hand joints in 3D space through a combined loss to generate 3D hand skeletons. The combined loss function $L_{total}$ is expressed as

$$L_{total} = \beta \times L_{3D} + \gamma \times L_{kine}, \tag{4.8}$$

where $\beta, \gamma$ is the weight corresponding to each loss. $L_{3D}$ is the 3D hand joint loss, which is represented by

$$L_{3D} = \sum_{i=0}^{20} ||h_i^{pred} - h_i^{gt}||_2, \tag{4.9}$$

where $h_i^{gt} = (x_i, y_i, z_i)$, $i = 0, 1, \ldots, 20$, is the ground truth, and $h_i^{pred}$ is the result of the $i$-th joint predicted by the proposed network.

$L_{kine}$ denotes the kinematic loss for hand motion, inspired by [14], which addresses finger bending in four distinct scenarios and imposes specific constraints for each. We simplify the relationship between finger joints into two geometric categories: collinear and coplanar, applying constraints for both. A human hand can be seen as a segmented rigid object. Each phalanx is treated as a rigid body, and these phalanges are connected via joints, enabling a range of hand movements. The notation $A$, $B$, $C$, and $D$ are used to represent three phalanges and the fingertip, with $A$ marking the base of the finger. Each joint is assigned its corresponding 3D coordinates. When the finger is extended, the four joints are collinear; when the finger is bent, the joints become non-collinear but remain coplanar. Figure 4.7 illustrates these two scenarios. The kinematic loss $L_{kine}$ can be expressed as $L_{kine} = \lambda L_{cop} + (1-\lambda) L_{col}$, where $\lambda$ is 1 in the collinear scenario and 0 in the coplanar one. $L_{col}$ represents the collinear loss, and $L_{cop}$ refers to the coplanar loss. For collinear cases, the length between the phalanges satisfies the condition

$$||B - A|| + ||C - B|| + ||D - C|| < (\phi + 1)||D - A||, \tag{4.10}$$

where $\phi$ is set to 0.01. Moreover, the angle between the vector corresponding to each phalanx and the direction vector $\mathbf{e}$ of the finger should remain small enough, so that $t < \cos(\overrightarrow{AB}, \mathbf{e}_d) < 1$. Here, $t$ is a value close to 1, set to 0.99 in *mmHand*. This condition also holds for $\overrightarrow{BC}$ and $\overrightarrow{CD}$.

## 4.5 Mesh Reconstruction

**Fig. 4.7** Two geometric relationships of finger joints

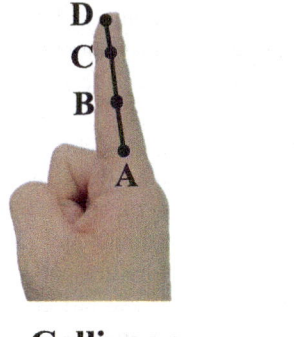
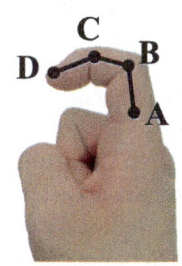

**Collinear**    **Coplanar**

Thus, the loss for collinear cases is formulated as:

$$L_{col} = \max(\|AB\| + \|BC\| + \|CD\| - 1.01\|AD\|, 0)$$

$$+ \max\left\{p - \frac{\vec{AB} \cdot \mathbf{e}_d}{\|\vec{AB}\|}, 0\right\}$$

$$+ \max\left\{p - \frac{\vec{BC} \cdot \mathbf{e}_d}{\|\vec{BC}\|}, 0\right\} \quad (4.11)$$

$$+ \max\left\{p - \frac{\vec{CD} \cdot \mathbf{e}_d}{\|\vec{CD}\|}, 0\right\}.$$

For coplanar cases, the direction vector of each phalanx is orthogonal to the normal vector $\mathbf{e}_n$ of the plane. Therefore, the coplanar loss is defined as:

$$L_{cop} = \vec{AB} \cdot \mathbf{e}_n + \vec{BC} \cdot \mathbf{e}_n + \vec{CD} \cdot \mathbf{e}_n. \quad (4.12)$$

Using the aggregated loss function, *mmHand* predicts the 3D coordinates of 21 hand joints through fully connected layers, which generates 3D hand skeletons.

## 4.5 Mesh Reconstruction

Once the 3D hand skeletons of 21 joints are obtained, *mmHand* next constructs detailed 3D hand meshes. Hand meshes provide a more intricate geometric representation, allowing for a more realistic depiction of hand gestures.

With advancements in 3D scanning technology, parametric hand models have been introduced to enhance the realism and precision of hand gesture reconstruction [15]. These models are grounded in the anatomical and kinematic properties of the human hand, representing it as a 3D entity controlled by specific parameters. By adjusting these parameters, the model can capture any conceivable hand shape or gesture. In *mmHand*, we utilize a hand model with articulated and non-rigid deformations (MANO) [16] to reconstruct 3D hand meshes. The MANO model is built upon the skinned multi-person linear model (SMPL) [17]. MANO addresses the intricate and flexible nature of human hands, using a mathematical formulation to represent finger movements and poses. More specifically, a differentiable function $M(\beta, \theta)$ leverages a set of shape parameters $\beta \in \mathbb{R}^{10}$ to adjust the hand's shape, and a set of pose parameters $\theta \in \mathbb{R}^{21 \times 3}$ to control its joint configurations. The shape parameters $\beta$ are derived from a principal component analysis of hand scans, while $\theta$ represents joint rotations in axis-angle form, expressed as

$$M(\beta, \theta) = W(T_p(\beta, \theta), J(\beta), \theta, \mathcal{W}), \tag{4.13}$$

where $W(\cdot)$ denotes a linear blend skinning function [18], $T_p$ represents the deformed template hand mesh, $J(\beta)$ gives the positions of the hand joints, and $\mathcal{W}$ refers to the skinning weights. The deformed template $T_p$ is obtained by applying the parameters to a base template, denoted as

$$T_p(\beta, \theta) = \vec{T} + B_s(\beta) + B_p(\theta), \tag{4.14}$$

where $\vec{T}$ is the standard template corresponding to the initial pose (T-pose) of the model, while $B_s(\beta)$ and $B_p(\theta)$ represent the shape and pose blend shapes, respectively.

To generate 3D hand meshes using the MANO model, we first calculate the pose parameters $\theta$ and the shape parameters $\beta$ independently. Next, the standard template $\vec{T}$ is modified using these parameters, as described in Eq. (4.14), to produce the final 3D hand meshes. The process of reconstructing hand meshes from the 21 hand joints $J_{3D}$ is illustrated in Fig. 4.8.

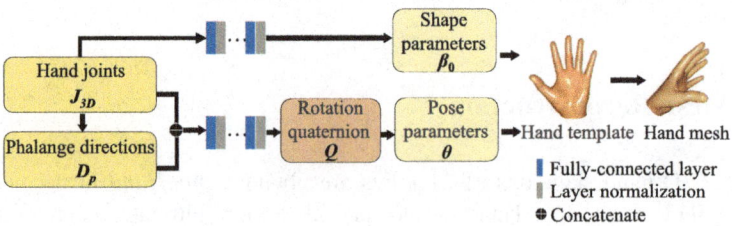

**Fig. 4.8** Hand mesh reconstruction

The distribution of hand skeleton joints in space reflects both the overall size and internal geometry of hand shapes. As a result, there exists a mapping between the reconstructed skeletons and the corresponding hand shapes. *mmHand* takes the reconstructed skeletons as input and utilizes three fully-connected layers with layer normalization to predict the shape parameters $\beta$, which are then used to generate the virtual hand shape. Afterward, the rotation parameters of the joints $\theta$ are inferred. Estimating the rotation of all hand joints, $\theta$, from the 3D skeleton is an inverse kinematics problem [19]. To address this problem in an end-to-end manner, *mmHand* leverages a deep learning framework to learn the relationship between joint coordinates and their corresponding rotations. Specifically, *mmHand* applies fully-connected layers with layer normalization to estimate the pose parameters $\theta$. The direction vectors of the phalanges $D_p \in \mathbb{R}^{20 \times 3}$ are calculated from the 3D coordinates of the 21 hand joints $J_{3D}$. These vectors $D_p$ and the coordinates $J_{3D}$ are then flattened and concatenated to serve as the input to the neural network. Providing explicit direction information for the phalanges enhances the accuracy of joint rotation predictions. To improve computational efficiency, the network outputs the rotation quaternions $\mathbf{Q} \in \mathbb{R}^{21 \times 4}$ for all joints. Finally, the quaternions $\mathbf{Q} \in \mathbb{R}^{21 \times 4}$ are converted into their corresponding axis-angle representations $\theta$.

The hand template $\vec{T}$ receives the pose parameters $\theta$ and shape parameters $\beta$ as inputs. Using the deformed template, *mmHand* constructs a 3D hand mesh from the 3D hand joints to realistically represent the hand's gesture, which realizes realistic and dynamic hand gesture reconstruction.

## 4.6 Evaluation

In this section, we perform experiments to assess the performance of *mmHand* in real-world conditions.

### 4.6.1 Evaluation Setup

*mmHand* is developed using a COTS mmWave radar (TI IWR1443 [20]) paired with a data acquisition card (TI DCA1000EVM [21]). The mmWave radar employs 3 transmit and 4 receive antennas to create a virtual antenna array leveraging TDM-MIMO technology. It transmits chirp signals in the frequency range of 77–81 GHz, with a chirp cycle time of 80 μs. Each chirp is sampled 64 times. In each frame, the 3 transmit antennas sequentially emit chirps for 64 cycles. TI mmWave Studio is set up on a desktop with an Intel Core i5-10440F processor to interface with the radar. The deep learning model is trained on an NVIDIA RTX 3090 Ti GPU. Ground truth data is obtained through a depth camera, and 21 hand joints are extracted using MediaPipe Hands [22] for the ground truth.

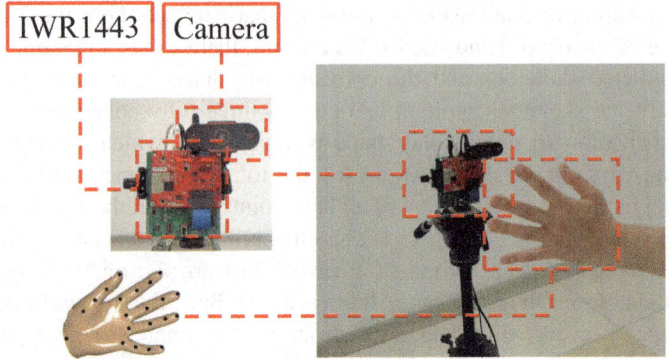

**Fig. 4.9** Experimental setup

Figure 4.9 illustrates the experimental setting where the mmWave radar and camera are positioned at the same location. Both devices are activated concurrently to gather mmWave signals and images, respectively. A total of 10 volunteers participated in the experiment including 5 males and 5 females. During data collection, the volunteers stood in front of the radar and positioned their hands at a distance of 20–40 cm from it. The gestures performed were continuous and included interaction gestures and counting gestures, which are common and spontaneous actions in everyday life. Both the mmWave radar and camera recorded the mmWave data and corresponding labels (i.e., coordinates of 21 hand joints). A total of 150,000 valid mmWave frames along with ground truth labels were collected from each participant. The experiment was conducted across three distinct environments: classrooms, corridors, and playgrounds. Additionally, to assess the robustness and performance of the model in unique scenarios, a small dataset was also gathered from volunteers wearing various gloves and holding different objects during testing.

We perform fivefold cross-validation during both the training and evaluation phases. Specifically, the data from 10 participants is partitioned into 5 subsets, with each subset containing data from 2 participants. In the $k$-th fold of cross-validation, the $k$-th subset is held out as the test set, while the remaining 4 subsets are combined to form the training set for model learning. This cross-validation procedure is designed to assess the model's performance while accounting for variations in both users and gestures. During training, the starting learning rate is set to 0.001 and follows a cosine decay schedule. The batch size is set to 16, and the model undergoes training for 500 epochs.

We employ the following evaluation metrics:

- *Mean Per Joint Position Error (MPJPE)* is the average Euclidean distance (*mm*) between the predicted hand joints and the ground truth, defined as

$$MPJPE = \frac{1}{N} \sum_{i=1}^{N} \left\| J_i^p - J_i^t \right\|, \tag{4.15}$$

## 4.6 Evaluation

where $N$ represents the total number of joints, $J_i^p$ is the predicted position of the $i$-th joint, and $J_i^t$ is the corresponding ground truth position.

- *Percentage of Correct Keypoints in 3D Space (3D-PCK)* refers to the proportion of correctly predicted hand joints based on various thresholds, and is calculated as

$$PCK_k = \frac{\sum_i \delta\left(\frac{d_i}{d} < T_k\right)}{\sum_i 1}, \quad (4.16)$$

where $T_k$ is a manually defined threshold, $d_i$ is the Euclidean distance between the predicted and ground truth position of the $i$-th joint, $d$ is a normalization factor, and $\delta$ is the indicator function.

- *Area Under the Curve (AUC)* refers to the area under the *3D-PCK* curve.

### 4.6.2 Overall Performance

We first evaluate the overall performance of *mmHand* in generating 3D hand skeletons and reconstructing 3D hand meshes. Figure 4.10 illustrates examples of hand skeletons and 3D meshes corresponding to various gestures. We can see that the 21 hand joints accurately represent different hand gestures. Also, the reconstructed 3D hand meshes show realistic animations that align with the user's hand gestures. *mmHand* is also capable of capturing and reconstructing dynamic hand gestures. Figure 4.11 presents examples of hand meshes for two consecutive gestures. The generated hand meshes clearly demonstrate the smooth transition between hand poses for continuous gesture sequences.

We further conduct a quantitative evaluation of *mmHand* by assessing MPJPE, 3D-PCK, and AUC for the 21 hand joints. Figures 4.12 and 4.13 show the MPJPE and 3D-PCK for each individual user, with the 3D-PCK threshold set at 40 mm.

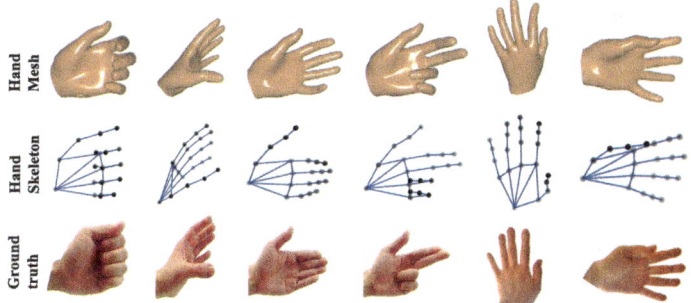

**Fig. 4.10** Examples of hand meshes and hand skeletons for different gestures

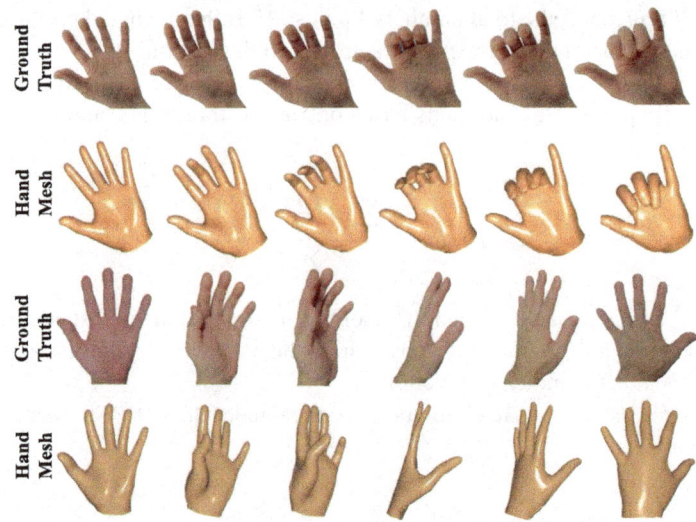

**Fig. 4.11** Examples of hand mesh reconstruction for continuous gestures

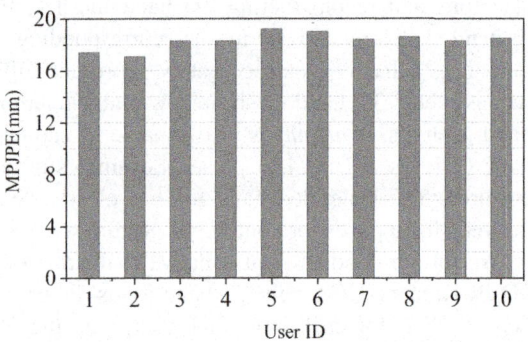

**Fig. 4.12** Per-participant MPJPE

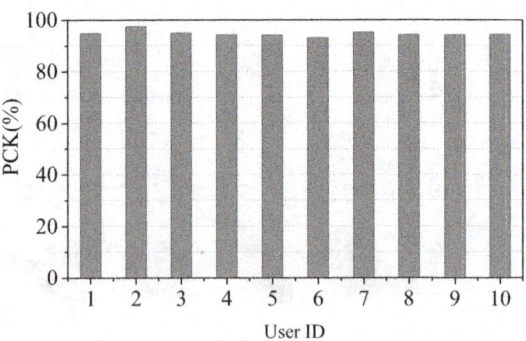

**Fig. 4.13** Per-participant 3D-PCK

## 4.6 Evaluation

Overall, *mmHand* achieves an average MPJPE of 18.3 mm and a 3D-PCK of 95.1%, with average standard deviations of 2.96 mm and 1.17%, respectively. These results demonstrate *mmHand*'s capability to accurately predict the 21 hand joints with small errors. Moreover, as seen in Figs. 4.12 and 4.13, the variations in MPJPE and 3D-PCK across users are negligible. For instance, the difference in MPJPE and 3D-PCK between user 2 (with the lowest MPJPE and highest 3D-PCK) and user 6 (with the highest MPJPE and lowest 3D-PCK) is just 2.9 mm and 3.3%, respectively. This underscores the robustness and effectiveness of *mmHand* in hand joint regression across various individuals.

To thoroughly evaluate *mmHand*'s performance in regressing various hand joints, we categorize the 21 joints into palm joints and finger joints, and subsequently calculate the average 3D-PCK and AUC across all users. Figure 4.14 illustrates the 3D-PCK performance of *mmHand*'s hand joint regression, with thresholds ranging from 0 to 60 mm. As seen, the 3D-PCK increases significantly as the threshold value rises. The overall 3D-PCK reaches 95.1% when the threshold is set to 40 mm. We also compute the AUC for the 3D-PCK curve, where a higher AUC signifies better performance. The results indicate that *mmHand* achieves an AUC of 0.707, demonstrating strong performance in hand joint regression. Figure 4.15 depicts the CDF of MPJPE across all hand joints. From the figure, we observe that 90.2% of the MPJPE values for the predicted hand joints fall within 30 mm. Furthermore, we note slight variations in the performance of joint regression across different regions of the hand. This discrepancy is likely due to the palm's relatively rigid structure during gestures, leading to more stable joint positions. In contrast, the fingers exhibit greater flexibility and are more prone to interactions, making the regression of finger joints more challenging.

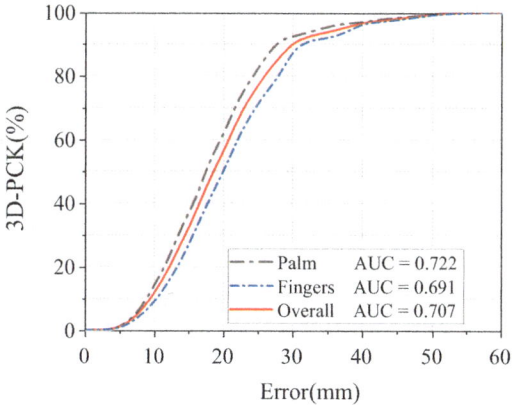

**Fig. 4.14** 3D-PCK under different error thresholds

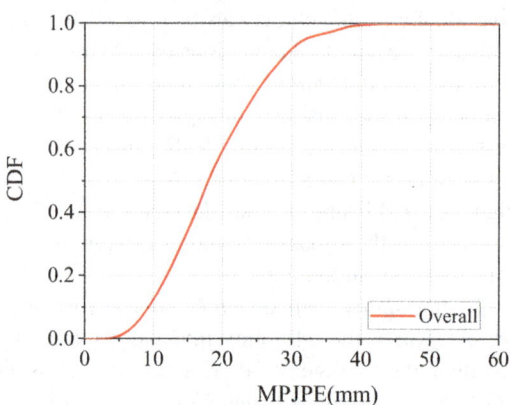

**Fig. 4.15** CDF of MPJPE

### 4.6.3 Comparison with Existing Methods

We evaluate the performance of *mmHand* by comparing it with several existing techniques, encompassing both vision-based and wireless signal-based approaches. The vision-based methods include Cascade [23], CrossingNet [24], DeepPrior++ [25], and HBE [26], which are evaluated by computing the MPJPE on two well-known 3D hand pose datasets, MSRA [23] and ICVL [27]. Given the challenge of constructing a one-to-one mmWave dataset corresponding to MSRA and ICVL for a direct comparison, we instead utilize our self-collected mmWave dataset and present a comparative analysis with these vision-based methods. For the wireless signal-based methods, we consider mm4Arm [11] and HandFi [28], where mm4Arm relies on mmWave signals, and HandFi uses WiFi signals. While these two wireless signal-based methods lack datasets that can be fully reproduced for comparison, we still align with their experimental configurations and collect mmWave data under similar conditions to ensure a reasonably fair comparison. Specifically, in a controlled lab environment, users perform the same hand gestures as described in the respective setups of the two studies [11, 28], while our mmWave radar continuously captures the corresponding signals. The collected mmWave data, along with the ground truth labels, are then used to compute the MPJPE, and these results are compared with the reported results from mm4Arm and HandFi.

Table 4.1 presents the MPJPE values for *mmHand* and six other existing methods. The comparison with the four vision-based methods is based on the MPJPE results from our own experiments, while the comparison with the two wireless signal-based techniques is based on the results obtained from data collected according to their experimental setups. We observe that although *mmHand*'s MPJPE slightly lags behind other advanced vision-based approaches due to the resolution constraints of mmWave signals, the difference is small. For example, the MPJPE difference between *mmHand* and the average value of 10.94 mm for the vision methods is less than 10 mm. When compared to the wireless signal-based methods, it is evident that mm4Arm shows better performance with self-collected data in terms of MPJPE than

## 4.6 Evaluation

**Table 4.1** MPJPE of *mmHand* and existing methods

| Methods | Dataset | MPJPE | mmHand |
|---|---|---|---|
| Cascade [23] | MSRA | 15.2 | 18.3 |
| | ICVL | 9.9 | |
| CrossingNet [24] | MSRA | 12.2 | |
| | ICVL | 10.2 | |
| DeepPrior++ [25] | MSRA | 9.5 | |
| HBE [26] | ICVL | 8.62 | |
| mm4Arm | Self-collected | 4.07 | 20.4 |
| HandFi | Self-collected | 20.7 | 19.0 |

our method. However, it necessitates users to maintain a specific orientation of their forearms facing the radar, which could hinder user experience in dynamic interaction contexts. On the other hand, *mmHand* demonstrates comparable performance to HandFi using its own collected data. These results show that *mmHand* is capable of performing efficient hand joint regression, comparable to both vision-based and wireless signal-based solutions.

### 4.6.4 Impact of Distance

The distance between the radar and the detected target can influence the quality of the signal reflection and lead to variations in signal patterns. Hence, distance plays a crucial role in affecting the user experience during hand gesture-based human-computer interaction. We evaluate the effect of the distance between the user's hand and the mmWave radar. In our experiments, the user's hand is positioned within a range of 20–40 cm for training the *mmHand* model. To investigate the influence of varying distances, we position the user's hand at distances ranging from 20 to 80 cm. Figures 4.16 and 4.17 display the MPJPE and 3D-PCK for hand joint regression

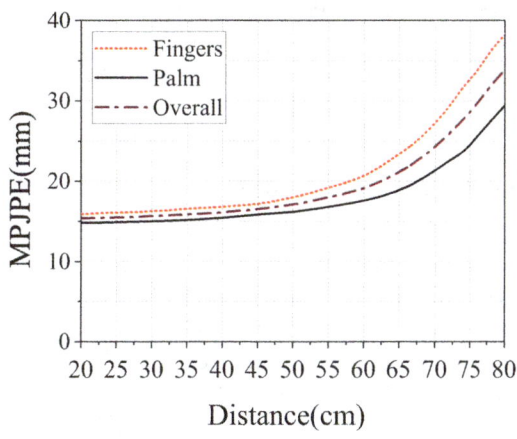

**Fig. 4.16** MPJPE in different distances

**Fig. 4.17** 3D-PCK in different distances

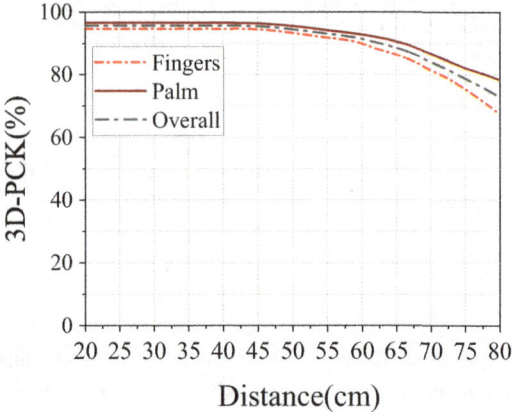

**Fig. 4.18** Experimental setup of different angles

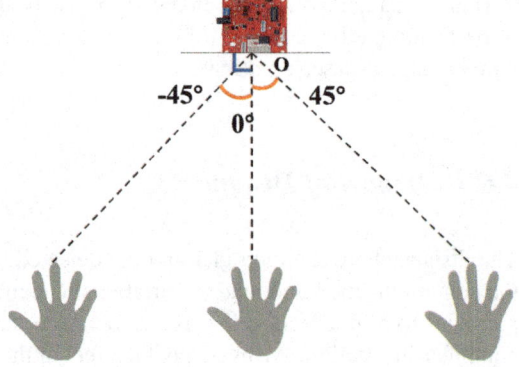

at different distances, with the 3D-PCK threshold set to 40 mm. From the results, we observe that both the MPJPE and 3D-PCK remain fairly consistent when the distance is between 20 and 60 cm. However, when the distance exceeds 60 cm, the MPJPE begins to increase while the 3D-PCK starts to decrease. Additionally, we notice that the MPJPE for the palm joints is generally lower than for the finger joints, and the 3D-PCK for the palm joints is higher, indicating that regression accuracy is greater for the palm joints compared to the finger joints at varying distances.

### 4.6.5 Impact of Angle

We evaluate the performance of *mmHand* when the user's hand is oriented at various angles towards the mmWave radar. In this experiment, the user's hand is placed at angles ranging from $-45°$ to $45°$, as depicted in Fig. 4.18. The angles have the gap around $15°$, and we quantitatively measure MPJPE and 3D-PCK at different angles. Figure 4.19 illustrates the MPJPE and 3D-PCK results across angles from $-45°$ to

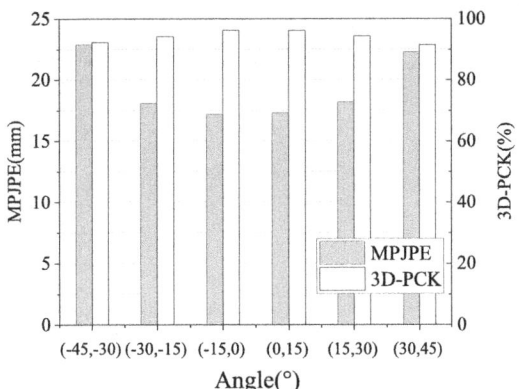

**Fig. 4.19** MPJPE and 3D-PCK in different angles

45°, with 15° steps. In this setup, the hand is positioned 40 cm away from the radar, and the 3D-PCK threshold is set to 40 mm. The results show that as the absolute value of the angle increases, the errors in hand joint regression tend to increase as well. When the angle exceeds 30°, there is a significant increase in the regression errors. This phenomenon is due to the decrease in angle estimation accuracy as the angle deviates further from the radar's line of sight. Despite these variations, when the angle remains within the range of $-30°$ to $+30°$, the average MPJPE and 3D-PCK values are 17.95 mm and 95.78%, respectively, indicating that *mmHand* can still effectively estimate hand joints and generate accurate hand gestures.

### 4.6.6 Impact of Human Body

When a user is positioned in front of a mmWave radar, the user's body may influence the propagation and reflection of mmWave signals, potentially leading to substantial effects on hand pose estimation. Therefore, we evaluate the impact of the human body on *mmHand*. Two distinct experimental setups are used where the user is positioned differently. In the first setup (type 1), the user stands directly in front of the radar with their hand extended forward, performing a variety of gestures. In the second setup (type 2), the user stands to the side of the radar, put out the hand in front of the radar. Figures 4.20 and 4.21 present the MPJPE and 3D-PCK results for each user in both experimental configurations. In the first setup, where the user is directly facing the radar, the overall MPJPE is 19.1 mm, and the overall 3D-PCK is 93.6%. In the second setup, where the user is positioned sideways to the radar, the MPJPE decreases slightly to 18.1 mm, and the 3D-PCK increases to 95.4%. The differences in performance between these two configurations are small. This is due to the fact that *mmHand* filters out the majority of the signals unrelated to the hand during the pre-processing stage, reducing the impact of the user's body position.

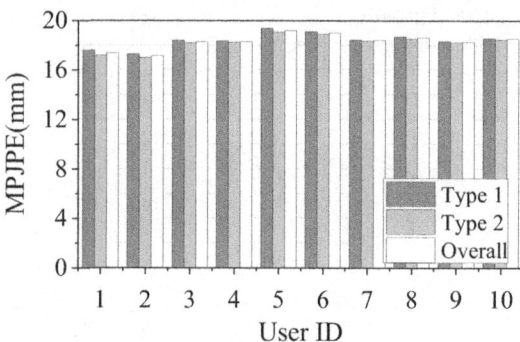

**Fig. 4.20** Per-participant MPJPE for type 1 and type 2

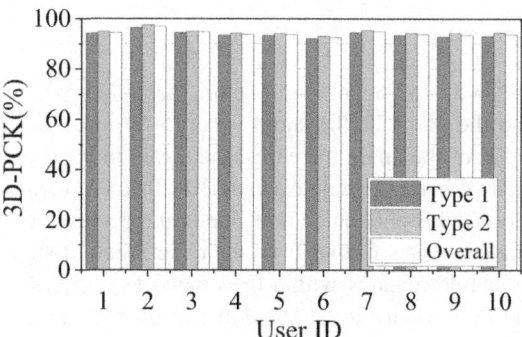

**Fig. 4.21** Per-participant PCK for type 1 and type 2

As a result, the hand pose estimation accuracy remains largely unaffected by the position of the human body.

## 4.6.7 Impact of Gloves

We perform experiments to evaluate the influence of different types of gloves on hand gesture reconstruction. In this experiment, users wear two gloves respectively: silk gloves and cotton gloves. The data collected under these conditions are directly used as the testing set to evaluate the performance in regressing the 21 hand joints. Figure 4.22 illustrates the hand joint regression results and the hand mesh reconstructions for users wearing each type of glove. From the results, we observe that *mmHand* accurately predicts the palm's position, but there are some inaccuracies in predicting the fingers, with some joints appearing to lean towards each other. The overall MPJPE for both types of gloves is 28.6 mm, and the 3D-PCK is 86.3%. Compared to the scenario without gloves, the accuracy of the hand joint regression decreases slightly when gloves are used. This is because the glove materials are also captured by the mmWave signals, causing some distortion in the perceived hand shape and results in minor deviations in the generated hand meshes.

4.6 Evaluation

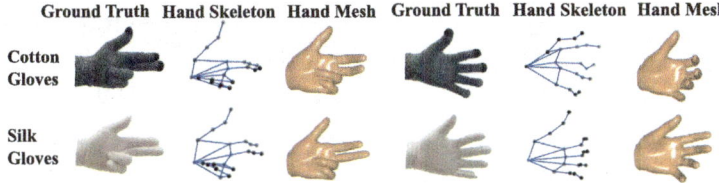

**Fig. 4.22** Examples of hand pose estimation when users wear gloves

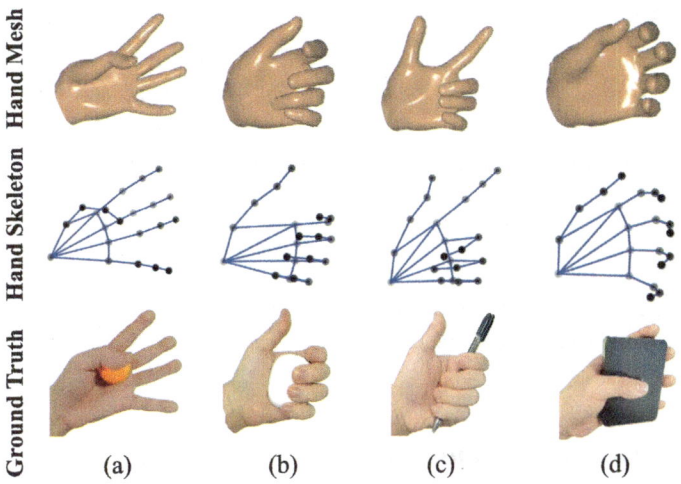

**Fig. 4.23** Examples of hand pose estimation when users hold different objects

### 4.6.8 Impact of Handheld Object

We evaluate the performance of *mmHand* when users are holding various objects. In this experiment, users hold four different objects: a table tennis ball, a headphone case, a pen, and a power bank. Figure 4.23 presents examples of *mmHand*'s ability to reconstruct 3D hand gestures while the user holds an object. The results in Fig. 4.23a and b show that when the held objects are small and primarily situated in the palm region, *mmHand* can accurately regress the 21 hand joints and reconstruct the corresponding 3D hand meshes. This is due to that these objects cause small interference with the reflected signals. Moreover, since the objects are centered in the hand, they predominantly affect the palm area, leaving the fingers largely unaffected, which allows for precise finger estimation. However, when the held object obstructs or distorts the signal reflections from the finger region, or if the object covers a large portion of the hand, *mmHand* may experience a drop in performance. For example, in Fig. 4.23c, *mmHand* incorrectly interprets the pen as a finger, while in Fig. 4.23d, the estimated fingers do not correspond well to the

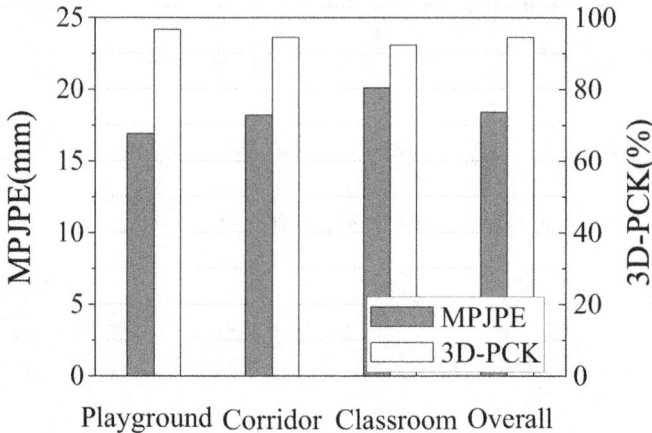

**Fig. 4.24** MPJPE and 3D-PCK in different environments

actual positions. This degradation is mainly caused by the interference caused by the reflected signals from the object.

### 4.6.9 Impact of Environment

To investigate the influence of background environments, we evaluate the performance of *mmHand* in three different environments: a playground, a corridor, and a classroom. The playground is an expansive, open space without obvious obstacles. The corridor has a static background with a few passers-by, while the classroom has a more complex static background along with more dynamic human movement. Figure 4.24 illustrates the MPJPE and 3D-PCK results across these three environments. The data shows that the variations across different environments are negligible. For example, the MPJPE difference between the playground and classroom is only 3.2 mm. This is because of *mmHand*'s ability to effectively localize the hand region through bandpass filtering on the mmWave signals, which filters out the background noise to sense hand gestures.

### 4.6.10 Impact of Obstacle

We evaluate the performance of *mmHand* in scenarios where obstacles are placed between the radar and the hand, blocking the direct line-of-sight propagation. This evaluation can reveal if *mmHand* can overcome the limitations of vision-based methods. In this experiment, we introduce A4 paper, a piece of cloth, and a thin

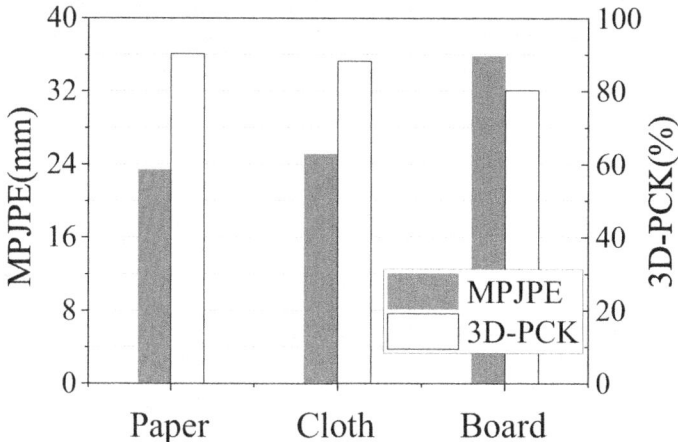

**Fig. 4.25** MPJPE and 3D-PCK under different occlusions

board as obstacles and calculate the MPJPE and 3D-PCK for the estimated hand joints in each scenario. Figure 4.25 presents the performance of MPJPE and 3D-PCK with different obstacles. The ground truth is derived from repeated gestures performed in line-of-sight conditions captured by cameras. It can be seen that different obstacles cause different levels of impact on hand gesture reconstruction. Specifically, the MPJPE for A4 paper and cloth are 23.4 and 25.1 mm, respectively, which represent slightly higher errors compared to the unobstructed cases. However, when a thin wood board is used as an obstacle, *mmHand*'s performance declines significantly, with an average of 35.8 mm MPJPE and 80.3% 3D-PCK. These results demonstrate that *mmHand* can still generate accurate hand gestures even when obstacles like paper and cloth block the line-of-sight. Thus, *mmHand* is a robust solution to illumination changes and non-line-of-sight conditions for gesture-based interactions.

## 4.7 Summary

In this chapter, we introduce *mmHand*, a hand gesture reconstruction system that utilizes a COTS mmWave radar to generate 3D hand skeletons and continuously reconstruct 3D hand meshes. To extract hand motion-related features from mmWave signals, we first design an attention-based hourglass network, mmSpaceNe, to extract multi-scale spatial features of the hand. Then, we integrate a LSTM neural network model to extract temporal features. After that, we design a combined loss for above neural network model and regress hand joints in 3D space to generate 3D hand skeletons, and finally reconstruct 3D hand meshes using the MANO model. Extensive experiments in real environments demonstrate the effectiveness of mmHand on hand gesture reconstruction.

# References

1. Wang, R.Y., Popović, J.: Real-time hand-tracking with a color glove. ACM Trans. Graph. **28**(3), 1–8 (2009)
2. Hu, F., He, P., Xu, S., Li, Y., Zhang, C.: Fingertrak: Continuous 3d hand pose tracking by deep learning hand silhouettes captured by miniature thermal cameras on wrist. Proc. ACM Interact. Mob. Wearable Ubiquitous Technol. **4**(2), 1–24 (2020)
3. Kulon, D., Wang, H., Güler, R.A., Bronstein, M., Zafeiriou, S.: Single image 3d hand reconstruction with mesh convolutions. Preprint (2019). arXiv:1905.01326
4. Xu, C., Cheng, L.: Efficient hand pose estimation from a single depth image. In: Proceedings of the IEEE ICCV' 13. Sydney (2013)
5. Chen, X., Liu, Y., Dong, Y., Zhang, X., Ma, C., Xiong, Y., Zhang, Y., Guo, X.: Mobrecon: Mobile-friendly hand mesh reconstruction from monocular image. In: Proceedings of the IEEE/CVF CVPR' 22. New Orleans (2022)
6. Chen, X., Liu, Y., Ma, C., Chang, J., Wang, H., Chen, T., Guo, X., Wan, P., Zheng, W.: Camera-space hand mesh recovery via semantic aggregation and adaptive 2d-1d registration. In: Proceedings of the IEEE/CVF CVPR' 21. Virtual Conference (2021)
7. Tang, X., Wang, T., Fu, C.W.: Towards accurate alignment in real-time 3d hand-mesh reconstruction. In: Proceedings of the IEEE/CVF ICCV' 21. Virtual Conference (2021)
8. Li, Y., Zhang, D., Chen, J., Wan, J., Zhang, D., Hu, Y., Sun, Q., Chen, Y.: Towards domain-independent and real-time gesture recognition using mmwave signal. IEEE Trans. Mob. Comput. **22**(12), 7355–7369 (2022)
9. Yu, J.T., Yen, L., Tseng, P.H.: mmwave radar-based hand gesture recognition using range-angle image. In: Proceedings of the IEEE VTC2020-Spring. Virtual Conference. IEEE (2020)
10. Ren, Y., Lu, J., Beletchi, A., Huang, Y., Karmanov, I., Fontijne, D., Patel, C., Xu, H.: Hand gesture recognition using 802.11 ad mmwave sensor in the mobile device. In: Proceedings of the IEEE WCNCW' 21. IEEE, Nanjing (2021)
11. Liu, Y., Zhang, S., Gowda, M., Nelakuditi, S.: Leveraging the properties of mmwave signals for 3d finger motion tracking for interactive iot applications. Proc. ACM Meas. Anal. Comput. Syst. **6**(3), 1–28 (2022)
12. Hu, J., Shen, L., Sun, G.: Squeeze-and-excitation networks. In: Proceedings of the IEEE CVPR' 18. Salt Lake City (2018)
13. Gao, Z., Xie, J., Wang, Q., Li, P.: Global second-order pooling convolutional networks. In: Proceedings of the IEEE/CVF CVPR' 19. Long Beach (2019)
14. Madadi, M., Escalera, S., Baró, X., Gonzalez, J.: End-to-end global to local cnn learning for hand pose recovery in depth data. Preprint (2017). arXiv:1705.09606
15. Yu, F., Zeng, L., Pan, D., Sui, X., Tang, J.: Evaluating the accuracy of hand models obtained from two 3d scanning techniques. Sci. Rep. **10**(1), 11875 (2020)
16. Romero, J., Tzionas, D., Black, M.J.: Embodied hands: Modeling and capturing hands and bodies together. Preprint (2022). arXiv:2201.02610
17. Loper, M., Mahmood, N., Romero, J., Pons-Moll, G., Black, M.J.: Smpl: A skinned multi-person linear model. ACM Trans. Graph. **34**(6), 1–16 (2015)
18. Kavan, L., Žára, J.: Spherical blend skinning: a real-time deformation of articulated models. In: Proceedings of the I3D' 05. Washington (2005)
19. Critchlow, A.J.: Introduction to Robotics. MacMillan Press Ltd., London, England (1985)
20. Instruments, T.: Iwr1443 single-chip 76-to 81-ghz mmwave sensor. IWR1443 datasheet, May (2017)

# References

21. Texas Instruments: Dca1000evm: Real-time Data-capture Adapter for Radar Sensing Evaluation Module (2020)
22. Zhang, F., Bazarevsky, V., Vakunov, A., Tkachenka, A., Sung, G., Chang, C.L., Grundmann, M.: Mediapipe hands: On-device real-time hand tracking. Preprint (2020). arXiv:2006.10214
23. Sun, X., Wei, Y., Liang, S., Tang, X., Sun, J.: Cascaded hand pose regression. In: Proceedings of the IEEE/CVF CVPR' 15. Boston (2015)
24. Wan, C., Probst, T., Van Gool, L., Yao, A.: Crossing nets: dual generative models with a shared latent space for hand pose estimation. In: Proceedings of the IEEE/CVF CVPR' 17. Hawaii (2017)
25. Oberweger, M., Lepetit, V.: Deepprior++: improving fast and accurate 3d hand pose estimation. In: Proceedings of the IEEE ICCV Workshops' 17. Venice (2017)
26. Zhou, Y., Lu, J., Du, K., Lin, X., Sun, Y., Ma, X.: Hbe: Hand branch ensemble network for real-time 3d hand pose estimation. In: Proceedings of the ECCV' 18. Munich (2018)
27. Tang, D., Jin Chang, H., Tejani, A., Kim, T.K.: Latent regression forest: Structured estimation of 3d articulated hand posture. In: Proceedings of the IEEE/CVF CVPR' 14. Columbus (2014)
28. Ji, S., Zhang, X., Zheng, Y., Li, M.: Construct 3d hand skeleton with commercial wifi. In: Proceedings of the ACM SenSys' 23, pp. 322–334 (2023)

# Chapter 5
# State-of-Art Research

**Abstract** In this chapter, we review state-of-art research related to the topic of the book. We first investigate human reconstruction approaches. Then, we introduce advanced mmWave radar sensing applications and research studies. Finally, we describe the work of mmWave-based human reconstruction research.

**Keywords** Human reconstruction technology · Vision approach · Wearable approach · mmWave sensing technology

## 5.1 Human Reconstruction Research

Human reconstruction has garnered significant interest due to its diverse applications in smart homes, motion-sensing games, virtual reality, etc. Aiming to estimate human poses, gestures, and even facial expressions through various sensing modalities, human reconstruction bridges the virtual worlds and physical worlds, allowing people to be digitally mapped to the real world. The sensing modalities can be broadly classified into vision-based methods, wearable device-based methods, and other emerging methods including RF sensing techniques.

### 5.1.1 Vision-based Approaches

Among existing approaches, vision-based human reconstruction is the most widely-used method due to the ubiquitous deployment and low cost of cameras. Vision-based methods rely on images and computer vision techniques to perform 2D and 3D human pose estimation tasks. Vision-based 2D human pose estimation usually aims to localize human body joints using images containing human. The approaches learn a mapping from the original image to the kinematic body model and generate joint coordinates [1]. Regression methods and heatmap-based methods are widely utilized to localize joint coordinates of human body, in which various deep neural networks play an important role [2]. Compared to single-person scenarios, multi-person human pose estimation requires identifying the number of individuals and associating keypoints with each person. This can be realized by either first detecting a single individual and further estimating a single pose [3], or detecting all body joints and then grouping them into different individuals [4]. The two approaches are known as top-down method and bottom-up method respectively. The key distinction between these two approaches is their computational efficiency. In top-down method, the processing time increases with the number of people in the image, as each detected individual requires separate pose estimation. Bottom-up method, on the other hand, tends to be more computationally efficient, as it does not require separate detections for each person, making it better suited for scenarios involving a large number of individuals.

3D human pose estimation aims to predict the locations of body joints in 3D space, which includes skeleton-only and mesh recovery. A large number of 3D human pose estimations are achieved by 2D-3D lifting, aiming to regress 3D joint locations from the estimated 2D joint locations [5]. Besides using 2D joint location, 3D human pose estimation can be also generated by 2D heatmaps [6, 7], which can avoid reconstruction ambiguity of over-reliance on the 2D pose detector. Considering that the human pose can be represented as a graph, graph convolutional networks (GCNs) have been applied to 2D to 3D pose enhancement problems [8, 9]. Additionally, many methods use prior knowledge based on kinematic models, such as bone-joint connection information, joint rotation properties, and fixed bone length ratios to estimate reasonable poses [10, 11]. The human mesh recovery method typically utilizes parametric body models, such as SMPL [12], to generate human meshes based on estimated joint locations. The vision-based approaches are popular due to the wide deployment of cameras. While they have achieved great success, they are vulnerable to light conditions and obstacles, which may limit the use scenarios in extensive IoT environments.

### 5.1.2 Wearable Device-Based Approaches

Wearable device-based methods utilize sensors, such as inertial measurement units (IMUs), electromyography (EMG) sensors, and depth sensors, to capture human motion and reconstruct body movements. These approaches are widely applied in healthcare, sports analysis, and rehabilitation due to their ability to provide continuous and real-time tracking. As Rashid and Hasan [13] categorize, four primary sensor types are employed for hand motion tracking.

Bend sensors are passive resistive devices designed to measure bending angles at hand joints. The thin, customizable form allows easy integration into glove knuckles, offering benefits like durability, low cost, and wide temperature adaptability. However, long-term use may cause substrate resistance drift due to material deformation, requiring frequent calibration. Commercial examples include the CyberGlove III [14], which uses conductive ink sensors to achieve sub-1° resolution and 120Hz sampling, and the 5DT Ultra [15] with optical bend sensors. Research efforts focus on optimizing sensor arrays [16] and developing novel materials [17] to enhance comfort and accuracy.

Stretch sensors are flexible devices that monitor body movements by measuring resistance changes proportional to deformation. They conform to joints and deformable body parts, offering comfort and dexterity, but face challenges in calibration consistency, slower response times, and material fatigue. Innovations include Lee et al.'s Ag NP/PDMS dual-strain sensor [18] and Bianchi's knitted piezoresistive fabric (KPF) [19], which tracks 19 degrees of freedom with five sensors. Glauser's dense stretch arrays [20] enable full-hand deformation tracking, though scalability and durability remain unresolved.

IMUs combine accelerometers, gyroscopes, and magnetometers to capture motion data via angular velocity and acceleration. They offer high sampling rates, low cost, and durability but may drift over time, necessitating recalibration. The Keyglove [21] uses IMUs for touch-based digital control, while the Hi5 VR Glove [22] employs 9-axis IMUs for high-precision VR motion capture. Research configurations, such as 17-axis systems [23], balance cost and performance, though long-term drift remains a limitation.

Magnetic sensors, such as Hall-effect devices, detect magnetic fields for contactless joint angle measurement. They provide low-cost, compact solutions with robust temperature performance but are susceptible to electromagnetic interference. The Humanglove [24] uses 20 Hall-effect sensors for precise finger tracking, while Finexus [25] employs fingertip electromagnets for spatial localization. Wu et al. [26] applied magnetic sensing in rehabilitation robotics for prosthetic hand control.

Despite their advantages, wearable systems face challenges such as hardware discomfort, sensor drift, and cost barriers, which hinder widespread adoption. Future research must address these limitations to improve user experience and accessibility.

## 5.2 mmWave Sensing Research

Millimeter wave sensing systems obtain information about the surrounding space by modulating electromagnetic waves to interact with the environment. This type of technology was originally designed to build wireless communication networks with large-capacity transmission, anti-interference capabilities, and ultra-low latency characteristics [27–29]. Its operating frequency band covers the range of 30–300 GHz. As technology evolves, the physical characteristics of this frequency band—including ultra-wide bandwidth, millimeter-level wavelengths, and compact antenna array design—give it environmental perception capabilities that exceed traditional communications.

A typical millimeter wave sensing device consists of a transmitting antenna array and a receiving antenna array to form a collaborative working system. The transmitting unit continuously radiates carefully designed modulated electromagnetic waves into space. These high-frequency signals interact with surrounding objects during propagation. When the electromagnetic waves are reflected by objects in the environment, the echo signals carrying the target characteristics are captured by the receiving array. By analyzing the phase, frequency, and intensity changes of the echo signals, the system can construct the three-dimensional spatial distribution of the target object and its dynamic motion trajectory, thereby realizing non-contact environmental monitoring.

### 5.2.1 mmWave Sensing in Autonomous Driving

Automotive applications using mmWave radars have become a pivotal research area, driven by their robust environmental sensing capabilities. Three key applications dominate: object detection, ego-motion estimation, and SLAM, each advancing through radar-only methods and sensor fusion.

Object detection focuses on identifying targets like vehicles and pedestrians. Radar signals are converted into point clouds or heatmaps for localization. Grid mapping [30, 31] and deep learning frameworks like the multi-scale grouping module (MSG) [32] highlight radar-only approaches. Fusion strategies, such as the nuScenes dataset [33], enable cross-modal systems like CameraRadarFusionNet (CRFNet) [34], which adapts fusion levels via Dropout learning, and spatial attention fusion (SAF) [35], merging radar and visual features. RODNet [36] uses radar spectra with cross-supervision from cameras, while milliEye [37] optimizes edge computing through decoupled architectures.

Ego-motion estimation estimates device motion using radar data. Landmark-based scan matching [38] and gradient-driven keypoint extraction [39] address noise and ghost reflections. Milli-RIO [40] combines radar with IMU for indoor six-degree-of-freedom estimation, while milliEgo [41] employs deep learning and cross-modal attention to handle sparse data. SAR imaging research [42, 43] further explores motion coherence for enhanced accuracy.

SLAM integrates mapping with localization. milliMap [44] uses cGANs with lidar supervision to generate dense maps, while radar SLAM [45] leverages radar cross-section (RCS) features and improved scan matching. Indoor systems exploit multipath effects [46] and dual-mode radar transmission [47], and sub-centimeter SLAM [48] combines AoA and time-of-arrival (ToA) estimation with 3D imaging. These advancements underscore mmWave radars' versatility in enabling autonomous systems through robust sensing and fusion.

### 5.2.2 mmWave Sensing in Smart Homes

The integration of mmWave radar modules into IoT devices has enabled a range of human-centric smart home applications, leveraging their ability to sense subtle movements and environmental interactions. Activity recognition serves as a foundational application, where mmWave signals classify human behaviors such as walking or typing by processing sparse point cloud data. Works like RadHAR [49] and m-Activity [50] utilize voxelization and graph neural networks to handle data sparsity, while multi-modal fusion approaches [51] combine point clouds with range-Doppler profiles to enhance accuracy. Deep learning architectures, including CNN-LSTM hybrids [49] and dual-view CNNs [52], extract spatial-temporal features for classification.

In speech recognition, mmWave systems capture vocal tract vibrations to recover speech content. WaveEar [53] and AmbiEar [54] use encoder-decoder networks to reconstruct high-fidelity audio, even leveraging indirect vibrations from surrounding objects. Generative models like cGAN [55] further enhance audio recovery from device vibrations. Vital sign monitoring focuses on non-invasive tracking of respiration and heartbeat. Techniques like phase unwrapping [56] and compressive sensing [57] isolate vital sign components, while multi-user systems such as mmVital [58] use beamforming to disentangle signals in dynamic environments.

User authentication leverages behavioral and biometric features. Gait-based methods like GaitCube [59] and CNN architectures [60] extract unique motion signatures, while biometric approaches such as HeartPrint [61] analyze heartbeat traits. Multi-modal fusion [62] and cross-modal adaptation [63] improve robustness across scenarios. Indoor positioning systems utilize FMCW-derived range and angle data. mmTrack [64] employs clustering and graph matching for trajectory estimation, while Kalman filter extensions [65, 66] address dynamic tracking challenges. The remaining two parts related to human reconstruction, namely pose estimation and gesture recognition, will be discussed in the Sect. 5.3.

### 5.2.3 mmWave Sensing in Industrial Manufacture

Millimeter-wave radars are increasingly integrated into industrial systems as versatile sensors, leveraging their advantages of cost-effectiveness, all-weather operation, and high-resolution sensing capabilities. These sensors are particularly valuable in scenarios where traditional methods like cameras or lidars face limitations, such as low visibility or material penetration requirements. Industrial applications are categorized into three main areas: imaging, measurement, and environmental monitoring.

In industrial imaging, mmWave radars enable spatial data extraction and image generation, even in adverse conditions. For instance, HawkEye [67] employs a conditional GAN to enhance fog-affected car imaging using low-resolution radar data, while a subsequent study [68] introduces a dual-generator architecture to produce 3D point clouds from radar signals. Hansen et al. [69] demonstrated high-resolution imaging of composite materials using custom radar setups, and Osprey [70] developed a tire wear assessment system by analyzing radar reflections from tire surfaces. These works highlight radar's adaptability for tasks ranging from underground mine inspections [71] to freehand scanning systems [72] that mitigate image artifacts through synthetic aperture techniques.

Industrial measurement applications exploit mmWave radars' precision in physical parameter detection. mmVib [73] introduced a vibration measurement system using IQ signal analysis, achieving micrometer-level accuracy. Ahmad et al. [74] designed a 60 GHz radar chipset for factory ranging and vibration monitoring, while studies on wind turbine blade clearance [75] and child presence detection [76] showcase radar's versatility in safety-critical scenarios. Security-related innovations include WaveSpy [77], which infers screen content via radar-based side-channel attacks, and distance-spoofing techniques [78] that exploit chirp modulation for automotive systems.

Environmental monitoring leverages mmWave radars to assess atmospheric and material conditions. ThermoWave [79] uses thermal scattering effects to measure temperature via paper tags, while studies [80–88] explore applications like gas identification, rainfall estimation, and soil moisture sensing. Insect detection also benefits from radar imaging [87, 88], demonstrating its role in ecological monitoring. These advancements underscore mmWave radars' potential to enhance industrial processes through robust, non-invasive sensing solutions.

## 5.3 mmWave-based Human Reconstruction

Due to the ability to operate in low-light conditions, penetrate obstacles, and preserve user privacy, researchers have explored mmWave sensing techniques for reconstructing human motion. The mmWave-based solution provides a non-contact and non-intrusive human reconstruction, which have yielded a number of research works recent days. This section discusses mmWave-based human reconstruction from two main aspects, i.e., pose estimation and gesture recognition.

### 5.3.1 mmWave-based Pose Estimation

Millimeter wave sensing-based pose estimation refers to the process of reconstructing a person's skeletal configuration by identifying the spatial coordinates of key body joints through the reflected mmWave signals. The technology captures subtle movements of the torso and limbs in non-contact manner, enabling applications such as motion analysis in sports, interactive gaming, and augmented reality. By leveraging deep learning, the micro motion sensitive mmWave signals are translated into skeletal joint coordinates.

A dominant approach in this filed is CNNs, which extract spatial features from radar data representations. For instance, mm-Pose [89] utilizes a split CNN architecture to map radar-imagery features to 17 joint coordinates, while Mars [90] employs 3D multi-channel inputs within a CNN framework for 3D joint prediction. Another study [91] integrates forward kinematics into a CNN to enhance stability for 25-joint estimation. Beyond CNNs, techniques inspired by natural language processing (NLP) have emerged, as seen in mmPose-NLP [92], which processes sequential radar point clouds via a GRU-based encoder-decoder, treating skeletal keypoints analogously to linguistic tokens. Similarly, transformer-based architectures like those in Wei et al. [93] refine positional accuracy by incorporating attention mechanisms and confidence networks.

Recent advancements address challenges in complex environments, such as multi-user scenarios and environmental variability. For example, a dual-branch network [94] combines global pose reconstruction with localized refinement to handle signal interference, while mPose [95] introduces domain adaptation to mitigate subject-specific and environmental biases. Multi-user tracking is realized [96], which employs spatial-temporal feature separation and ConvLSTM-based models to estimate and track joints across individuals. SynMotion [97] further extends this by synthesizing signals for few-shot skeleton tracking. Beyond skeletal estimation, human mesh reconstruction has gained traction. mmMesh [98] infers 3D meshes from sparse point clouds by aligning body parts and extrapolating occlusions, whereas M4mesh [99] addresses multi-user occlusion via temporal coherence. Hybrid approaches, such as fusing mmWave with visual data [100], enhance robustness in dynamic conditions.

## 5.3.2 mmWave-based Gesture Recognition

Gesture recognition is a critical application in human-computer interaction. By leveraging mmWave radar to interpret human motions through non-contact sensing, it captures hand-reflected signals to classify gestures or estimate hand positions. As a foundation for smart environments and device control, mmWave-based gesture recognition enables intuitive interactions by translating gestures into commands. While, some challenges posed by real-world variability, such as diverse environments, user orientations, and subject-specific motion patterns, are being extensively studied by researchers to expand the application scenario.

Research efforts to enhance robustness in complex scenarios have explored two primary directions. The first focuses on signal processing techniques that extract invariant features across varying conditions. For instance, mHomeGes [101] introduces user discovery to isolate target gestures from background noise, improving resilience to positional and orientation shifts. M-Gesture [102] employs pseudo-representative models to encode gesture trajectories and skeletal changes, enabling person-independent recognition. Temporal space-velocity spectrograms [103] further unify multi-modal gesture data, while DIGesture [104] synthesizes training samples by modeling correlations between signal variations and gesture dynamics, reducing dependency on extensive data collection. Techniques like the GreBsmo algorithm [105] filter static interference to isolate dynamic gestures, and intrinsic spectrogram transformations [106] standardize motion patterns across spatial and velocity variations via virtual coordinate systems. Frequency-ratio features [107] additionally mitigate unintended gesture interference, reinforcing robustness.

The second direction harnesses deep learning to automate feature extraction and classification. Early approaches, such as CNN-LSTM hybrids [108], demonstrated cross-user adaptability for dynamic gestures, while 3D CNNs [109] and Deep CNN architectures [110] learned embedded spatial-temporal patterns. Innovations like mmASL [111] integrate domain knowledge from sign language recognition through multi-task learning, enhancing environmental generalization. Graph-based models, such as temporal k-NN combined with self-attention MPNNs [112], process motion point clouds for gesture dynamics, whereas PointNet++-LSTM hybrid [113] exploits sparse 3D point clouds for real-time recognition. Transfer learning [114] and semi-supervised methods [115] further reduce annotation burdens by adapting models to new domains with limited labeled data.

Beyond classification, hand tracking extends functionality by quantifying motion trajectories for applications like VR control or virtual keyboards. Soli [116] models phase shifts in mmWave signals to infer hand displacement, which fuses instantaneous and dynamic phase data for location-agnostic tracking. Forearm muscle activation patterns [117] are alternatively analyzed to estimate hand movements indirectly. Deep learning further enhances precision, as seen in FCNN-based super-resolution localization [118] for continuous motion capture. Similarly, finger and object tracking leverages phase analysis [119] or channel impulse responses [120] to isolate targets and reconstruct trajectories, enabling applications like passive pen tracking during handwriting.

## 5.4 Summary of Existing Research

Existing research on human reconstruction faces inherent trade-offs between accuracy, user convenience, and practical deployment. Vision-based methods dominate due to camera ubiquity but remain constrained by environmental dependencies and privacy concerns. Wearable devices, such as IMUs and stretch sensors, provide continuous tracking but introduce hardware discomfort and calibration challenges. Innovations in sensor materials and configurations aim to balance accuracy and user experience, but are still limited by deployment costs.

The emerging mmWave radar-based techniques offer promising alternatives by enabling contact-free sensing through electromagnetic wave reflections. These methods perform well in poor illuminations and privacy-sensitive settings, but face signal processing complexity and multi-user occlusion challenges. Recent advances in mmWave human reconstruction focus on pose estimation and gesture recognition, which processes radar point clouds to reconstruct skeletal joints with various deep learning and domain adaptation methods. Multi-user reconstruction is also considered with techniques such as spatial-temporal separation to support real-world multi-user scenarios. Despite human body reconstruction, gesture recognition leverages invariant feature engineering and deep learning to achieve robust gesture classification, though environmental variability and signal interference remain hurdles.

Despite the progress, mmWave-based methods still require solutions for more fine-grained facial reconstruction and high-precision hand tracking, where subtle movements demand enhanced resolution and noise mitigation. Addressing ambiguity and improving availability is highly required for mmWave radar-based human reconstruction. For example, supporting scalable multi-user scenarios using commercial hardware is practical for real-world scenarios. Besides, improving generalization across environments and integrating physical models to address ambiguity in occluded regions are also remaining technical issues. For future work, to balance robustness and user-centric design, the convergence of advanced signal processing, lightweight neural architectures, and cross-modal fusion will drive next-generation human reconstruction technologies.

# References

1. Toshev, A., Szegedy, C.: Deeppose: Human pose estimation via deep neural networks. In: Proceedings of the IEEE Conference on Computer Vision and Pattern Recognition, pp. 1653–1660 (2014)
2. Zheng, C., Wu, W., Chen, C., Yang, T., Zhu, S., Shen, J., Kehtarnavaz, N., Shah, M.: Deep learning-based human pose estimation: a survey. ACM Comput. Surv. **56**(1), 1–37 (2023)
3. Cai, Y., Wang, Z., Luo, Z., Yin, B., Du, A., Wang, H., Zhang, X., Zhou, X., Zhou, E., Sun, J.: Learning delicate local representations for multi-person pose estimation. In: Computer Vision–ECCV 2020: 16th European Conference, Glasgow, UK, August 23–28, 2020, Proceedings, Part III 16, pp. 455–472. Springer (2020)
4. Pishchulin, L., Insafutdinov, E., Tang, S., Andres, B., Andriluka, M., Gehler, P.V., Schiele, B.: Deepcut: Joint subset partition and labeling for multi person pose estimation. In: Proceedings of the IEEE Conference on Computer Vision and Pattern Recognition, pp. 4929–4937 (2016)
5. Martinez, J., Hossain, R., Romero, J., Little, J.J.: A simple yet effective baseline for 3d human pose estimation. In: Proceedings of the IEEE International Conference on Computer Vision, pp. 2640–2649 (2017)
6. Tekin, B., Márquez-Neila, P., Salzmann, M., Fua, P.: Learning to fuse 2d and 3d image cues for monocular body pose estimation. In: Proceedings of the IEEE International Conference on Computer Vision, pp. 3941–3950 (2017)
7. Zhou, K., Han, X., Jiang, N., Jia, K., Lu, J.: Hemlets pose: Learning part-centric heatmap triplets for accurate 3d human pose estimation. In: Proceedings of the IEEE/CVF International Conference on Computer Vision, pp. 2344–2353 (2019)
8. Choi, H., Moon, G., Lee, K.M.: Pose2mesh: Graph convolutional network for 3d human pose and mesh recovery from a 2d human pose. In: Computer Vision–ECCV 2020: 16th European Conference, Glasgow, UK, August 23–28, 2020, Proceedings, Part VII 16, pp. 769–787. Springer (2020)
9. Zhao, L., Peng, X., Tian, Y., Kapadia, M., Metaxas, D.N.: Semantic graph convolutional networks for 3d human pose regression. In: Proceedings of the IEEE/CVF Conference on Computer Vision and Pattern Recognition, pp. 3425–3435 (2019)
10. Georgakis, G., Li, R., Karanam, S., Chen, T., Košecká, J., Wu, Z.: Hierarchical kinematic human mesh recovery. In: Computer Vision–ECCV 2020: 16th European Conference, Glasgow, UK, August 23–28, 2020, Proceedings, Part XVII 16, pp. 768–784. Springer (2020)

11. Kundu, J.N., Seth, S., Rahul, M., Rakesh, M., Radhakrishnan, V.B., Chakraborty, A.: Kinematic-structure-preserved representation for unsupervised 3d human pose estimation. In: Proceedings of the AAAI Conference on Artificial Intelligence, vol. 34, pp. 11312–11319 (2020)
12. Loper, M., Mahmood, N., Romero, J., Pons-Moll, G., Black, M.J.: Smpl: A skinned multi-person linear model. In: Seminal Graphics Papers: Pushing the Boundaries, vol. 2, pp. 851–866 (2023)
13. Rashid, A., Hasan, O.: Wearable technologies for hand joints monitoring for rehabilitation: a survey. Microelectron. J. **88**, 173–183 (2019)
14. CyberGlove Systems: Cyberglove iii (2020). Available online: http://www.cyberglovesystems.com/cyberglove-iii
15. 5DT Inc.: 5dt Data Glove Ultra Series (2020). Available online: http://www.5dt.com/downloads/dataglove/ultra/5DTDataGloveUltraDatasheet.pdf
16. Saggio, G.: A novel array of flex sensors for a goniometric glove. Sensors Actuators A Phys. **205**, 119–125 (2014)
17. Shen, Z., Yi, J., Li, X., Lo, M.H.P., Chen, M.Z., Hu, Y., Wang, Z.: A soft stretchable bending sensor and data glove applications. Robot. Biomimetics **3**(1), 22 (2016)
18. Lee, J., Kim, S., Lee, J., Yang, D., Park, B.C., Ryu, S., Park, I.: A stretchable strain sensor based on a metal nanoparticle thin film for human motion detection. Nanoscale **6**(20), 11932–11939 (2014)
19. Bianchi, M., Haschke, R., Büscher, G., Ciotti, S., Carbonaro, N., Tognetti, A.: A multi-modal sensing glove for human manual-interaction studies. Electronics **5**(3), 42 (2016)
20. Glauser, O., Panozzo, D., Hilliges, O., Sorkine-Hornung, O.: Deformation capture via soft and stretchable sensor arrays. ACM Trans. Graph. **38**(2), 1–16 (2019)
21. Yang, C.C., Hsu, Y.L.: A review of accelerometry-based wearable motion detectors for physical activity monitoring. Sensors **10**(8), 7772–7788 (2010)
22. O'Flynn, B., Sanchez, J.T., Connolly, J., Condell, J., Curran, K., Gardiner, P., Downes, B.: Integrated smart glove for hand motion monitoring. In: The Sixth International Conference on Sensor Device Technologies and Applications. International Academy, Research, and Industry Association (2015)
23. Hsiao, P.C., Yang, S.Y., Lin, B.S., Lee, I.J., Chou, W.: Data glove embedded with 9-axis imu and force sensing sensors for evaluation of hand function. In: 2015 37th Annual International Conference of the IEEE Engineering in Medicine and Biology Society (EMBC), pp. 4631–4634. IEEE (2015)
24. Humanware: The humanglove (2023). http://www.hmw.it/en/humanglove.html
25. Chen, K.Y., Patel, S.N., Keller, S.: Finexus: Tracking precise motions of multiple fingertips using magnetic sensing. In: Proceedings of the 2016 CHI Conference on Human Factors in Computing Systems, pp. 1504–1514 (2016)
26. Wu, J., Huang, J., Wang, Y., Xing, K.: Rlsesn-based pid adaptive control for a novel wearable rehabilitation robotic hand driven by pm-ts actuators. Int. J. Intell. Comput. Cybern. **5**(1), 91–110 (2012)
27. Yu, Q., Han, C., Bai, L., Choi, J., Shen, X.: Low-complexity multiuser detection in millimeter-wave systems based on opportunistic hybrid beamforming. IEEE Trans. Veh. Technol. **67**(10), 10129–10133 (2018)
28. Qiao, J., Shen, X.S., Mark, J.W., Shen, Q., He, Y., Lei, L.: Enabling device-to-device communications in millimeter-wave 5g cellular networks. IEEE Commun. Mag. **53**(1), 209–215 (2015)
29. He, S., Zhang, Y., Wang, J., Zhang, J., Ren, J., Zhang, Y., Zhuang, W., Shen, X.: A survey of millimeter-wave communication: Physical-layer technology specifications and enabling transmission technologies. Proc. IEEE **109**(10), 1666–1705 (2021)
30. Prophet, R., Li, G., Sturm, C., Vossiek, M.: Semantic segmentation on automotive radar maps. In: 2019 IEEE Intelligent Vehicles Symposium (IV), pp. 756–763. IEEE (2019)
31. Lombacher, J., Laudt, K., Hahn, M., Dickmann, J., Wöhler, C.: Semantic radar grids. In: 2017 IEEE Intelligent Vehicles Symposium (IV), pp. 1170–1175. IEEE (2017)

# References

32. Schumann, O., Hahn, M., Dickmann, J., Wöhler, C.: Semantic segmentation on radar point clouds. In: 2018 21st International Conference on Information Fusion (FUSION), pp. 2179–2186. IEEE (2018)
33. Caesar, H., Bankiti, V., Lang, A.H., Vora, S., Liong, V.E., Xu, Q., Krishnan, A., Pan, Y., Baldan, G., Beijbom, O.: nuscenes: A multimodal dataset for autonomous driving. In: Proceedings of the IEEE/CVF Conference on Computer Vision and Pattern Recognition, pp. 11621–11631 (2020)
34. Nobis, F., Geisslinger, M., Weber, M., Betz, J., Lienkamp, M.: A deep learning-based radar and camera sensor fusion architecture for object detection. In: 2019 Sensor Data Fusion: Trends, Solutions, Applications (SDF), pp. 1–7. IEEE (2019)
35. Chang, S., Zhang, Y., Zhang, F., Zhao, X., Huang, S., Feng, Z., Wei, Z.: Spatial attention fusion for obstacle detection using mmwave radar and vision sensor. Sensors **20**(4), 956 (2020)
36. Wang, Y., Jiang, Z., Li, Y., Hwang, J.N., Xing, G., Liu, H.: Rodnet: A real-time radar object detection network cross-supervised by camera-radar fused object 3d localization. IEEE J. Sel. Top. Signal Process. **15**(4), 954–967 (2021)
37. Shuai, X., Shen, Y., Tang, Y., Shi, S., Ji, L., Xing, G.: millieye: A lightweight mmwave radar and camera fusion system for robust object detection. In: Proceedings of the International Conference on Internet-of-Things Design and Implementation, pp. 145–157 (2021)
38. Cen, S.H., Newman, P.: Precise ego-motion estimation with millimeter-wave radar under diverse and challenging conditions. In: 2018 IEEE International Conference on Robotics and Automation (ICRA), pp. 6045–6052. IEEE (2018)
39. Cen, S.H., Newman, P.: Radar-only ego-motion estimation in difficult settings via graph matching. In: 2019 International Conference on Robotics and Automation (ICRA), pp. 298–304. IEEE (2019)
40. Almalioglu, Y., Turan, M., Lu, C.X., Trigoni, N., Markham, A.: Milli-rio: ego-motion estimation with low-cost millimetre-wave radar. IEEE Sensors J. **21**(3), 3314–3323 (2020)
41. Lu, C.X., Saputra, M.R.U., Zhao, P., Almalioglu, Y., De Gusmao, P.P., Chen, C., Sun, K., Trigoni, N., Markham, A.: milliego: single-chip mmwave radar aided egomotion estimation via deep sensor fusion. In: Proceedings of the 18th Conference on Embedded Networked Sensor Systems, pp. 109–122 (2020)
42. Gao, X., Roy, S., Xing, G.: Mimo-sar: A hierarchical high-resolution imaging algorithm for mmwave fmcw radar in autonomous driving. IEEE Trans. Veh. Technol. **70**(8), 7322–7334 (2021)
43. Steiner, M., Grebner, T., Waldschmidt, C.: Millimeter-wave sar-imaging with radar networks based on radar self-localization. IEEE Trans. Microwave Theory Tech. **68**(11), 4652–4661 (2020)
44. Lu, C.X., Rosa, S., Zhao, P., Wang, B., Chen, C., Stankovic, J.A., Trigoni, N., Markham, A.: See through smoke: robust indoor mapping with low-cost mmwave radar. In: Proceedings of the 18th International Conference on Mobile Systems, Applications, and Services, pp. 14–27 (2020)
45. Li, Y., Liu, Y., Wang, Y., Lin, Y., Shen, W.: The millimeter-wave radar slam assisted by the rcs feature of the target and imu. Sensors **20**(18), 5421 (2020)
46. Hao, Z., Yan, H., Dang, X., Ma, Z., Jin, P., Ke, W.: Millimeter-wave radar localization using indoor multipath effect. Sensors **22**(15), 5671 (2022)
47. Lee, S., Kwon, S.Y., Kim, B.J., Lim, H.S., Lee, J.E.: Dual-mode radar sensor for indoor environment mapping. Sensors **21**(7), 2469 (2021)
48. Aladsani, M., Alkhateeb, A., Trichopoulos, G.C.: Leveraging mmwave imaging and communications for simultaneous localization and mapping. In: ICASSP 2019-2019 IEEE International Conference on Acoustics, Speech and Signal Processing (ICASSP), pp. 4539–4543. IEEE (2019)
49. Singh, A.D., Sandha, S.S., Garcia, L., Srivastava, M.: Radhar: Human activity recognition from point clouds generated through a millimeter-wave radar. In: Proceedings of the 3rd ACM Workshop on Millimeter-wave Networks and Sensing Systems, pp. 51–56 (2019)

50. Wang, Y., Liu, H., Cui, K., Zhou, A., Li, W., Ma, H.: m-activity: Accurate and real-time human activity recognition via millimeter wave radar. In: ICASSP 2021–2021 IEEE International Conference on Acoustics, Speech and Signal Processing (ICASSP), pp. 8298–8302. IEEE (2021)
51. Huang, Y., Li, W., Dou, Z., Zou, W., Zhang, A., Li, Z.: Activity recognition based on millimeter-wave radar by fusing point cloud and range–doppler information. Signals **3**(2), 266–283 (2022)
52. Yu, C., Xu, Z., Yan, K., Chien, Y.R., Fang, S.H., Wu, H.C.: Noninvasive human activity recognition using millimeter-wave radar. IEEE Syst. J. **16**(2), 3036–3047 (2022)
53. Xu, C., Li, Z., Zhang, H., Rathore, A.S., Li, H., Song, C., Wang, K., Xu, W.: Waveear: Exploring a mmwave-based noise-resistant speech sensing for voice-user interface. In: Proceedings of the 17th Annual International Conference on Mobile Systems, Applications, and Services, pp. 14–26 (2019)
54. Zhang, J., Zhou, Y., Xi, R., Li, S., Guo, J., He, Y.: Ambiear: mmwave based voice recognition in nlos scenarios. Proc. ACM Interact. Mob. Wearable Ubiquitous Technol. **6**(3), 1–25 (2022)
55. Hu, P., Ma, Y., Santhalingam, P.S., Pathak, P.H., Cheng, X.: Milliear: Millimeter-wave acoustic eavesdropping with unconstrained vocabulary. In: IEEE INFOCOM 2022-IEEE Conference on Computer Communications, pp. 11–20. IEEE (2022)
56. Alizadeh, M., Shaker, G., De Almeida, J.C.M., Morita, P.P., Safavi-Naeini, S.: Remote monitoring of human vital signs using mm-wave fmcw radar. IEEE Access **7**, 54958–54968 (2019)
57. Wang, Y., Wang, W., Zhou, M., Ren, A., Tian, Z.: Remote monitoring of human vital signs based on 77-ghz mm-wave fmcw radar. Sensors **20**(10), 2999 (2020)
58. Yang, Z., Pathak, P.H., Zeng, Y., Liran, X., Mohapatra, P.: Vital sign and sleep monitoring using millimeter wave. ACM Trans. Sensor Netw. **13**(2), 1–32 (2017)
59. Ozturk, M.Z., Wu, C., Wang, B., Liu, K.R.: Gaitcube: Deep data cube learning for human recognition with millimeter-wave radio. IEEE Int. Things J. **9**(1), 546–557 (2021)
60. Jiang, X., Zhang, Y., Yang, Q., Deng, B., Wang, H.: Millimeter-wave array radar-based human gait recognition using multi-channel three-dimensional convolutional neural network. Sensors **20**(19), 5466 (2020)
61. Wang, Y., Gu, T., Luan, T.H., Lyu, M., Li, Y.: Heartprint: Exploring a heartbeat-based multiuser authentication with single mmwave radar. IEEE Int. Things J. **9**(24), 25324–25336 (2022)
62. Cao, D., Liu, R., Li, H., Wang, S., Jiang, W., Lu, C.X.: Cross vision-rf gait re-identification with low-cost rgb-d cameras and mmwave radars. Proc. ACM Interact. Mobile Wearable Ubiquitous Technol. **6**(3), 1–25 (2022)
63. Xu, W., Song, W., Liu, J., Liu, Y., Cui, X., Zheng, Y., Han, J., Wang, X., Ren, K.: Mask does not matter: Anti-spoofing face authentication using mmwave without on-site registration. In: Proceedings of the 28th Annual International Conference on Mobile Computing and Networking, pp. 310–323 (2022)
64. Wu, C., Zhang, F., Wang, B., Liu, K.R.: mmtrack: Passive multi-person localization using commodity millimeter wave radio. In: IEEE INFOCOM 2020-IEEE Conference on Computer Communications, pp. 2400–2409. IEEE (2020)
65. Jiang, M., Guo, S., Luo, H., Cui, G.: Continuous tracking of indoor human targets based on millimeter wave radar. In: 2022 Asia-Pacific Signal and Information Processing Association Annual Summit and Conference (APSIPA ASC), pp. 2071–2076. IEEE (2022)
66. Wang, X., Zhang, Z., Zhao, N., Zhang, Y., Huang, D.: Indoor localization and trajectory tracking system based on millimeter-wave radar sensor. In: 2021 IEEE 10th Data Driven Control and Learning Systems Conference (DDCLS), pp. 1141–1147. IEEE (2021)
67. Guan, J., Madani, S., Jog, S., Gupta, S., Hassanieh, H.: Through fog high-resolution imaging using millimeter wave radar. In: Proceedings of the IEEE/CVF Conference on Computer Vision and Pattern Recognition, pp. 11464–11473 (2020)

# References

68. Sun, Y., Huang, Z., Zhang, H., Cao, Z., Xu, D.: 3drimr: 3d reconstruction and imaging via mmwave radar based on deep learning. In: 2021 IEEE International Performance, Computing, and Communications Conference (IPCCC), pp. 1–8. IEEE (2021)
69. Hansen, S., Bredendiek, C., Briese, G., Froehly, A., Herschel, R., Pohl, N.: A sige-chip-based d-band fmcw-radar sensor with 53-ghz tuning range for high resolution measurements in industrial applications. IEEE Trans. Microwave Theory Tech. **70**(1), 719–731 (2021)
70. Prabhakara, A., Singh, V., Kumar, S., Rowe, A.: Osprey: A mmwave approach to tire wear sensing. In: Proceedings of the 18th International Conference on Mobile Systems, Applications, and Services, pp. 28–41 (2020)
71. Brooker, G., Hennessey, R., Lobsey, C., Bishop, M., Widzyk-Capehart, E.: Seeing through dust and water vapor: Millimeter wave radar sensors for mining applications. J. Field Robot. **24**(7), 527–557 (2007)
72. Alvarez-Narciandi, G., López-Portugués, M., Las-Heras, F., Laviada, J.: Freehand, agile, and high-resolution imaging with compact mm-wave radar. IEEE Access **7**, 95516–95526 (2019)
73. Jiang, C., Guo, J., He, Y., Jin, M., Li, S., Liu, Y.: mmvib: micrometer-level vibration measurement with mmwave radar. In: Proceedings of the 26th Annual International Conference on Mobile Computing and Networking, pp. 1–13 (2020)
74. Ahmad, W.A., Wessel, J., Ng, H.J., Kissinger, D.: Iot-ready millimeter-wave radar sensors. In: 2020 IEEE Global Conference on Artificial Intelligence and Internet of Things (GCAIoT), pp. 1–5. IEEE (2020)
75. Zhang, L., Wei, J.: Measurement and control method of clearance between wind turbine tower and blade-tip based on millimeter-wave radar sensor. Mech. Syst. Signal Process. **149**, 107319 (2021)
76. Caddemi, A., Cardillo, E.: Automotive anti-abandon systems: a millimeter-wave radar sensor for the detection of child presence. In: 2019 14th International Conference on Advanced Technologies, Systems and Services in Telecommunications (TELSIKS), pp. 94–97. IEEE (2019)
77. Li, Z., Ma, F., Rathore, A.S., Yang, Z., Chen, B., Su, L., Xu, W.: Wavespy: Remote and through-wall screen attack via mmwave sensing. In: 2020 IEEE Symposium on Security and Privacy (SP), pp. 217–232. IEEE (2020)
78. Miura, N., Machida, T., Matsuda, K., Nagata, M., Nashimoto, S., Suzuki, D.: A low-cost replica-based distance-spoofing attack on mmwave fmcw radar. In: Proceedings of the 3rd ACM Workshop on Attacks and Solutions in Hardware Security Workshop, pp. 95–100 (2019)
79. Chen, B., Li, H., Li, Z., Chen, X., Xu, C., Xu, W.: Thermowave: a new paradigm of wireless passive temperature monitoring via mmwave sensing. In: Proceedings of the 26th Annual International Conference on Mobile Computing and Networking, pp. 1–14 (2020)
80. Mead, J.B., Pazmany, A.L., Sekelsky, S.M., McIntosh, R.E.: Millimeter-wave radars for remotely sensing clouds and precipitation. Proc. IEEE **82**(12), 1891–1906 (1994)
81. Hattenhorst, B., Piotrowsky, L., Pohl, N., Musch, T.: An mmwave sensor for real-time monitoring of gases based on real refractive index. IEEE Trans. Microwave Theory Tech. **69**(11), 5033–5044 (2021)
82. Cao, D., Lin, Y., Ren, G., Gao, Y., Dong, W.: Mmliquid: Liquid identification using mmwave. In: China Conference on Wireless Sensor Networks, pp. 1–18. Springer (2022)
83. Dai, Q., Huang, Y., Wang, L., Ruby, R., Wu, K.: mm-humidity: Fine-grained humidity sensing with millimeter wave signals. In: 2018 IEEE 24th International Conference on Parallel and Distributed Systems (ICPADS), pp. 204–211. IEEE (2018)
84. Han, C., Huo, J., Gao, Q., Su, G., Wang, H.: Rainfall monitoring based on next-generation millimeter-wave backhaul technologies in a dense urban environment. Remote Sensing **12**(6), 1045 (2020)
85. Golovachev, Y., Etinger, A., Pinhasi, G., Pinhasi, Y.: The effect of weather conditions on millimeter wave propagation. Int. J. Circuits Syst. Signal Process. **13**, 690–695 (2019)

86. Chen, W., Feng, Y., Cardamis, M., Jiang, C., Song, W., Ghannoum, O., Hu, W.: Soil moisture sensing with mmwave radar. In: Proceedings of the 6th ACM Workshop on Millimeter-Wave and Terahertz Networks and Sensing Systems, pp. 19–24 (2022)
87. Tahir, N., Brooker, G.: Toward the development of millimeter wave harmonic sensors for tracking small insects. IEEE Sensors J. **15**(10), 5669–5676 (2015)
88. Sheikh, F., Prokscha, A., Batra, A., Lessy, D., Salah, B., Sievert, B., Degen, M., Rennings, A., Jalali, M., Svejda, J.T., et al.: Towards continuous real-time plant and insect monitoring by miniaturized thz systems. IEEE J. Microwaves **3**(3), 913–937 (2023)
89. Sengupta, A., Jin, F., Zhang, R., Cao, S.: mm-pose: Real-time human skeletal posture estimation using mmwave radars and cnns. IEEE Sensors J. **20**(17), 10032–10044 (2020)
90. An, S., Ogras, U.Y.: Mars: mmwave-based assistive rehabilitation system for smart healthcare. ACM Trans. Embed. Comput. Syst. **20**(5s), 1–22 (2021)
91. Hu, S., Sengupta, A., Cao, S.: Stabilizing skeletal pose estimation using mmwave radar via dynamic model and filtering. In: 2022 IEEE-EMBS International Conference on Biomedical and Health Informatics (BHI), pp. 1–6. IEEE (2022)
92. Sengupta, A., Cao, S.: mmpose-nlp: a natural language processing approach to precise skeletal pose estimation using mmwave radars. IEEE Trans. Neural Netw. Learn. Syst. **34**(11), 8418–8429 (2022)
93. Wei, G., Cui, C., Dong, X.: A transformer-based network for human pose estimation using millimeter wave radar data. In: 2023 International Applied Computational Electromagnetics Society Symposium (ACES-China), pp. 1–4. IEEE (2023)
94. Cao, Z., Ding, W., Chen, R., Zhang, J., Guo, X., Wang, G.: A joint global–local network for human pose estimation with millimeter wave radar. IEEE Int. Things J. **10**(1), 434–446 (2022)
95. Shi, C., Lu, L., Liu, J., Wang, Y., Chen, Y., Yu, J.: mpose: Environment-and subject-agnostic 3d skeleton posture reconstruction leveraging a single mmwave device. Smart Health **23**, 100228 (2022)
96. Kong, H., Xu, X., Yu, J., Chen, Q., Ma, C., Chen, Y., Chen, Y.C., Kong, L.: m3track: mmwave-based multi-user 3d posture tracking. In: Proceedings of the 20th Annual International Conference on Mobile Systems, Applications and Services, pp. 491–503 (2022)
97. Zhang, X., Li, Z., Zhang, J.: Synthesized millimeter-waves for human motion sensing. In: Proceedings of the 20th ACM Conference on Embedded Networked Sensor Systems, pp. 377–390 (2022)
98. Xue, H., Ju, Y., Miao, C., Wang, Y., Wang, S., Zhang, A., Su, L.: mmmesh: Towards 3d real-time dynamic human mesh construction using millimeter-wave. In: Proceedings of the 19th Annual International Conference on Mobile Systems, Applications, and Services, pp. 269–282 (2021)
99. Xue, H., Cao, Q., Ju, Y., Hu, H., Wang, H., Zhang, A., Su, L.: M4esh: mmwave-based 3d human mesh construction for multiple subjects. In: Proceedings of the 20th ACM Conference on Embedded Networked Sensor Systems, pp. 391–406 (2022)
100. Ding, H., Chen, Z., Zhao, C., Wang, F., Wang, G., Xi, W., Zhao, J.: Mi-mesh: 3d human mesh construction by fusing image and millimeter wave. Proc. ACM Interact. Mobile Wearable Ubiquitous Technol. **7**(1), 1–24 (2023)
101. Liu, H., Wang, Y., Zhou, A., He, H., Wang, W., Wang, K., Pan, P., Lu, Y., Liu, L., Ma, H.: Real-time arm gesture recognition in smart home scenarios via millimeter wave sensing. Proc. ACM Interact. Mobile Wearable Ubiquitous Technol. **4**(4), 1–28 (2020)
102. Liu, H., Zhou, A., Dong, Z., Sun, Y., Zhang, J., Liu, L., Ma, H., Liu, J., Yang, N.: M-gesture: Person-independent real-time in-air gesture recognition using commodity millimeter wave radar. IEEE Int. Things J. **9**(5), 3397–3415 (2021)
103. Zhang, K., Lan, S., Zhang, G.: On the effect of training convolution neural network for millimeter-wave radar-based hand gesture recognition. Sensors **21**(1), 259 (2021)
104. Li, Y., Zhang, D., Chen, J., Wan, J., Zhang, D., Hu, Y., Sun, Q., Chen, Y.: Towards domain-independent and real-time gesture recognition using mmwave signal. IEEE Trans. Mobile Comput. **22**(12), 7355–7369 (2022)

# References

105. Zhao, Y., Sark, V., Krstic, M., Grass, E.: Novel approach for gesture recognition using mmwave fmcw radar. In: 2022 IEEE 95th Vehicular Technology Conference (VTC2022-Spring), pp. 1–6. IEEE (2022)
106. Wu, J., Wang, J., Gao, Q., Cheng, M., Pan, M., Zhang, H.: Toward robust device-free gesture recognition based on intrinsic spectrogram of mmwave signals. IEEE Int. Things J. **9**(19), 19318–19329 (2022)
107. Liu, C., Li, Y., Ao, D., Tian, H.: Spectrum-based hand gesture recognition using millimeter-wave radar parameter measurements. IEEE Access **7**, 79147–79158 (2019)
108. Wang, S., Song, J., Lien, J., Poupyrev, I., Hilliges, O.: Interacting with soli: Exploring fine-grained dynamic gesture recognition in the radio-frequency spectrum. In: Proceedings of the 29th Annual Symposium on User Interface Software and Technology, pp. 851–860 (2016)
109. Hazra, S., Santra, A.: Short-range radar-based gesture recognition system using 3d cnn with triplet loss. IEEE Access **7**, 125623–125633 (2019)
110. Smith, J.W., Thiagarajan, S., Willis, R., Makris, Y., Torlak, M.: Improved static hand gesture classification on deep convolutional neural networks using novel sterile training technique. IEEE Access **9**, 10893–10902 (2021)
111. Santhalingam, P.S., Hosain, A.A., Zhang, D., Pathak, P., Rangwala, H., Kushalnagar, R.: mmasl: Environment-independent asl gesture recognition using 60 ghz millimeter-wave signals. Proc. ACM Interact. Mobile Wearable Ubiquitous Technol. **4**(1), 1–30 (2020)
112. Salami, D., Hasibi, R., Palipana, S., Popovski, P., Michoel, T., Sigg, S.: Tesla-rapture: a lightweight gesture recognition system from mmwave radar sparse point clouds. IEEE Trans. Mobile Comput. **22**(8), 4946–4960 (2022)
113. Palipana, S., Salami, D., Leiva, L.A., Sigg, S.: Pantomime: Mid-air gesture recognition with sparse millimeter-wave radar point clouds. Proc. ACM Interact. Mobile Wearable Ubiquitous Technol. **5**(1), 1–27 (2021)
114. Liu, H., Cui, K., Hu, K., Wang, Y., Zhou, A., Liu, L., Ma, H.: mtranssee: Enabling environment-independent mmwave sensing based gesture recognition via transfer learning. Proc. ACM Interact. Mobile Wearable Ubiquitous Technol. **6**(1), 1–28 (2022)
115. Yan, B., Wang, P., Du, L., Chen, X., Fang, Z., Wu, Y.: mmgesture: Semi-supervised gesture recognition system using mmwave radar. Expert Syst. Appl. **213**, 119042 (2023)
116. Lien, J., Gillian, N., Karagozler, M.E., Amihood, P., Schwesig, C., Olson, E., Raja, H., Poupyrev, I.: Soli: ubiquitous gesture sensing with millimeter wave radar. ACM Trans. Graph. **35**(4), 1–19 (2016)
117. Liu, Y., Zhang, S., Gowda, M., Nelakuditi, S.: Leveraging the properties of mmwave signals for 3d finger motion tracking for interactive iot applications. Proc. ACM Meas. Anal. Comput. Syst. **6**(3), 1–28 (2022)
118. Smith, J.W., Furxhi, O., Torlak, M.: An fcnn-based super-resolution mmwave radar framework for contactless musical instrument interface. IEEE Trans. Multimedia **24**, 2315–2328 (2021)
119. Wei, T., Zhang, X.: mtrack: High-precision passive tracking using millimeter wave radios. In: Proceedings of the 21st Annual International Conference on Mobile Computing and Networking, pp. 117–129 (2015)
120. Regani, S.D., Wu, C., Wang, B., Wu, M., Liu, K.R.: mmwrite: Passive handwriting tracking using a single millimeter-wave radio. IEEE Int. Things J. **8**(17), 13291–13305 (2021)

# Chapter 6
# Conclusions and Future Research Directions

**Abstract** In this chapter, we conclude the monograph and outline several promising research directions for future work.

**Keywords** mmWave sensing techniques · Few-shot learning · Cross domain · Integrated sensing and communication · Large language models

## 6.1 Conclusions

As smart homes increasingly adopt IoT devices, the boundary between the physical and digital worlds is narrowing. Human reconstruction is key to synchronizing human states in the cyber world, enabling applications like VR, AR, the metaverse, etc. Traditional approaches rely on wearables and cameras. Wearable-based approaches are intrusive and costly, while vision-based methods are sensitive to lighting conditions and raise privacy concerns. To overcome these challenges, millimeter wave sensing has emerged as a robust, privacy-preserving alternative. The technique is suited for non-intrusive human reconstruction, from full-body postures to detailed facial expressions and hand gestures. In this monograph, we make use of mmWave radar sensing techniques in human reconstruction. The main highlights are as follows.

In Chap. 1, we have introduced human reconstruction by outlining the core concepts, application scenarios, and commonly used approaches. We have then provided an overview of mmWave sensing technologies, including FMCW principles and parameter estimation methods, followed by a discussion of their emerging applications.

In Chap. 2, we have proposed a system that utilizes a single mmWave radar to simultaneously reconstruct and track the 3D postures of multiple users as they move, walk, or sit. It first isolates individual users from radar signals in multi-user environments. Then, it extracts each user's shape and motion features and employs a customized deep learning model to reconstruct their 3D postures. These postures are mapped into real-world 3D space, where a coordinate-corrected tracking method ensures accurate position tracking. Extensive experiments in real-world scenarios confirm the system's accuracy and robustness, highlighting its effectiveness for practical multi-user 3D posture tracking.

In Chap. 3, we have developed a passive and privacy-conscious framework for 3D facial expression reconstruction using a single mmWave radar. The framework processes pre-captured radar signals to extract key facial geometric features using a ConvNeXt model enhanced with triple loss embedding, capturing subtle facial dynamics. It then reconstructs robust, distance- and orientation-invariant facial shapes by applying a region-divided affine transformation to 68 facial landmarks. To further enhance expression fidelity, the system employs a region-based amplification technique before generating 3D avatars that reflect dynamic facial expressions. Real-world evaluations with 15 participants show that the system achieves accurate facial landmark tracking, demonstrating its effectiveness and practicality for real-world applications.

In Chap. 4, we have introduced a novel system that leverages mmWave signals for 3D hand pose reconstruction. The system can generate both 3D hand skeletons and detailed hand meshes. It begins by detecting the hand using mmWave signals and processing the radar data. Then, an attention-based hourglass network is proposed to extract multi-scale spatial features, while a long short-term memory network captures temporal dynamics. These features are used to regress 3D hand joint positions and generate accurate hand skeletons. Finally, a MANO-based model reconstructs continuous 3D hand meshes, accounting for both articulation and deformation. Extensive real-world experiments demonstrate that the system achieves accurate hand pose estimation and hand mesh reconstruction.

In Chap. 5, we have presented existing research relevant to human reconstruction. We have explored recent advancements in mmWave radar sensing technologies and their diverse applications. Finally, we have highlighted key research efforts focused on mmWave-based human reconstruction.

## 6.2 Future Research Directions

With the growing deployment of intelligent systems in smart homes, healthcare, industrial automation, and the metaverse, there is an increasing need for human reconstruction technologies. Millimeter wave radar, with its ability to sense fine-grained human movements under occlusion and in varying lighting conditions, has emerged as a promising solution. Based on the foundation laid by this monograph, several future research directions can be envisioned.

**Model Construction with Low Training Cost** Existing mmWave-based human reconstruction systems often rely on deep learning frameworks that demand large volumes of annotated data for supervised training. However, collecting and labeling high-quality mmWave data for diverse human poses, gestures, and facial expressions remains labor-intensive and expensive. To address this challenge, future research could explore different model training strategies. For example, few-shot and transfer learning methods can be utilized to adapt pre-trained models to new tasks or environments with minimal labeled data. Cross-modal training strategy leverages existing large-scale vision datasets to guide mmWave model training via domain adaptation or modality transformation. Self-supervised learning framework that leverages inherent signal structure to learn feature representations without extensive labels is another solution. Also, to enhance generalization in unseen scenarios, researchers can generate synthetic data using physics-based simulators to augment real data. These approaches will help reduce the cost of model development and accelerate the deployment of mmWave sensing applications in the real world.

**Extension to Diverse Environments and Configurations** Most existing mmWave human reconstruction systems are designed for controlled indoor scenarios with limited numbers of users. However, future applications will span more diverse environments. For example, crowded public spaces, where accurate multi-person tracking and posture reconstruction remain challenging due to interference and signal overlap. Dynamic outdoor environments, where environmental noise, mobility, and material scattering complicate signal interpretation. Besides, heterogeneous device platforms, where varying antenna layouts, radar specifications, and computing resources necessitate adaptable and generalizable solutions. Research can explore how to dynamically adapt reconstruction pipelines to such complex conditions, potentially using federated learning, edge AI, and collaborative sensing techniques to coordinate among multiple mmWave devices.

**Toward Integrated Sensing and Communication** As mmWave technologies evolve with 5G and upcoming 6G developments, the line between sensing and communication is increasingly blurred. mmWave technologies can simultaneously support high-bandwidth communication and fine-grained environmental perception. This convergence leads to many opportunities. For example, integrated sensing and communication (ISAC) systems that jointly optimize radar sensing and communication channel utilization. Resource-aware reconstruction algorithms can adaptively allocate bandwidth and computation depending on the sensing precision needed. Context-aware sensing, where user behavior or application context (e.g., VR immersion level, industrial task complexity) influences sensing granularity and frequency. Future research can explore how to embed human reconstruction capabilities within ISAC frameworks to support intelligent, low-latency, and continuous perception in resource-constrained environments.

**Integration of Large Language Models with mmWave Sensing** The rapid advancements in large language models (LLMs), exemplified by systems like ChatGPT, have demonstrated exceptional capabilities in natural language understanding, knowledge reasoning, and data generation. When combined with mmWave sensing, LLMs offer a promising new research direction that has the potential to transform how sensing data is interpreted, synthesized, and applied. One compelling application of LLMs lies in data synthesis. Given that mmWave sensing systems often require large-scale labeled data for training, collecting diverse, high-quality datasets is a key challenge. Leveraging the generative capabilities of LLMs, especially when paired with foundation models or multi-modal training frameworks, can alleviate this burden. Beyond data generation, LLMs offer a novel perspective on understanding and reasoning about physical-world interactions. Traditional mmWave sensing systems typically rely on neural networks that predict structured outputs (e.g., positions, labels, gestures) from raw radar data. However, these outputs often lack semantic richness or contextual interpretation. In contrast, LLMs, when paired with mmWave features or multi-modal embeddings, can infer higher-level insights from sensing signals—such as interpreting user intent, describing ongoing activities in natural language, or correlating patterns across time and space. Furthermore, the fusion of LLMs and mmWave sensing aligns well with multi-modal learning. When mmWave data is combined with other modalities (e.g., images, audio, inertial signals), LLMs can serve as integrative reasoning engines that unify diverse sensor streams and offer coherent, human-like explanations. This multi-modal reasoning could greatly enhance the reliability and adaptability of mmWave-based human reconstruction systems in real-world environments.

Made in the USA
Monee, IL
03 May 2026

49438397R00077